Biofueled Reciprocating Internal Combustion Engines

Biofueled Reciprocating Internal Combustion Engines

K.A. Subramanian

CRC Press

Taylor & Francis Group
Boca Raton   London   New York

CRC Press is an imprint of the
Taylor & Francis Group, an **informa** business

CRC Press
Taylor & Francis Group
6000 Broken Sound Parkway NW, Suite 300
Boca Raton, FL 33487-2742

International Standard Book Number-13: 978-1-138-03318-4 (Hardback)
International Standard Book Number-13: 978-1-138-74654-1 (Paperback)

Library of Congress Cataloging-in-Publication Data

Names: Subramanian, K. A., 1971- author.
Title: Biofueled reciprocating internal combustion engines / K.A. Subramanian.
Description: Boca Raton : CRC Press, 2017. | Includes bibliographical references and index.
Identifiers: LCCN 2017012992| ISBN 9781138033184 (hardback : acid-free paper)
| ISBN 9781315116785 (ebook)
Subjects: LCSH: Internal combustion engines. | Biomass energy.
Classification: LCC TJ756 .S83 2017 | DDC 621.43--dc23
LC record available at https://lccn.loc.gov/2017012992

Visit the Taylor & Francis Web site at
http://www.taylorandfrancis.com

and the CRC Press Web site at
http://www.crcpress.com

Contents

Preface

I warmly welcome you to read this book entitled *Biofueled Reciprocating Internal Combustion Engines*. I am thankful to you for sparing your valuable time to read this book. I am confident that you could find the important information pertaining to the area of biofuels and their utilization in internal combustion engines. The following questions are generally being asked by academic researchers, industrial researchers, students/scholars, policy makers, environmentalists, and ecologists who attend biofuels-related conferences, workshops, symposiums, seminars, lecture classes, research project meetings, and so forth.

- The question "Would a biofuels program affect the environment due to a decrease in forest cover by increasing land for energy crops?" is generally asked by environmentalists.
- The question "Would a biofuels program damage an ecosystem due to a decrease in habitat area and food web by increasing land for energy crops?" is generally asked by ecologists.
- The question "Would a biofuels program threaten food security due to land occupied by energy crops that would lead to a decrease in agricultural land?" is generally asked by socialists, the general public, and governments.
- The question "Are biofuels economical and price competitive with fossil fuels?" is generally asked by economists.
- The question "Would biofuels be beneficial in terms of performance improvement and emissions reduction in internal combustion engines?" is generally asked by researchers.
- The question "What viable biofuels are practical to implement in vehicle fleets?" is generally asked by original engine manufacturers (OEMs) and the oil industries.
- The question "Would a biofuels program enhance energy security, a sustainable environment, and a sustainable ecosystem?" is generally asked by policy makers and government bureaucrats.

Even though the answers to the above questions are complex, some are available in the literature but are too widely scattered and not properly linked to one another. This book attempts to bring the information together in one place. From looking at the comprehensive literature survey carried out for this book, one can see that the information related to biofuels and their utilization in internal combustion engines is scattered and therefore needed to be compiled in a coherent manner. In addition, the relationship between biofuels and the environment and ecology has never been properly examined. This book also sets out to highlight the need for biofuel as a way to strengthen and maintain sustainable energy, the environment, and ecosystems.

Chapter 1, "Introduction to Biofuels," provides an overview of the basics of energy flow, demand-supply gap of fossil fuels, different types of transportation modes (inter- and intra-modes), applications of internal combustion engines, air pollution from internal combustion engines, the potential of biofuel resources, the impact of bioenergy on ecosystems and the environment, comparison of bioenergy with other renewable energy, first- and second-generation biofuels, carbon-neutral fuel, and biofuel policy around the world. The impacts of biofuels on energy, the environment, and ecosystems are clearly explained in this chapter.

Chapter 2, "Production of Biofuels," provides information related to biofuel resources and production technologies. The main suitable biofuels (methanol, ethanol, butanol, biogas, hydrogen, biodiesel, Fischer-Tropsch diesel, and dimethyl ether) for internal combustion engines are identified. The important fuel qualities of the selected biofuels are discussed in Chapter 3, "Biofuel Quality for Internal Combustion Engines," and the effects of fuel quality on engine performance and emissions are also discussed. Chapter 4, "Introduction to Internal Combustion Engines," reviews the first and second laws of thermodynamics, thermodynamic cycles (Otto, diesel, and dual cycles), introduces two- and four-stroke engines, provides brief information about premixed and homogeneous charged engines (homogeneous charge compression ignition, premixed charge compression ignition, and reactivity controlled charge compression ignition), introduces variable speed automotive engines and constant speed power genset engines, provides a comparison of internal combustion engines with other types of power plants and a comparison between spark-ignition and compression-ignition engines, NOx emission reduction methods, and solved numerical problems.

In Chapter 5, "Basic Processes of Internal Combustion Engines," the important process of spark-ignition engines, including fuel induction and injection methods, air flow (swirl and tumble-type charge flow), fuel spray characteristics, mixing, ignition, combustion, and emissions are highlighted. Relevant model and correlations equations pertaining to engine processes are also compiled in this chapter. Chapter 6, "Utilization of Biofuels in Spark-Ignition Engines," focuses on the effect of biofuels on performance and engine emissions. Shortcomings, such as power drop, backfire, and certain emissions including aldehyde at higher-level emissions are discussed. Chapter 7, "Utilization of Biofuels in Compression-Ignition Engines," provides comprehensive information on the effects of biodiesel, Fischer-Tropsch diesel, and dimethyl ether on injection, fuel spray, and performance and emissions characteristics of compression-ignition engines. Technical issues, including higher fuel injection pressure and spray penetration distance, automatic advancement of injection timing with biodiesel, and wall impingement with biodiesel, are discussed in detail. Technological solutions to these problems are also highlighted in this chapter.

In Chapter 8, "Biofueled Reactivity Controlled Compression-Ignition Engines," advanced technologies such as homogeneous charge compression ignition, premixed charge compression ignition, and reactivity controlled compression ignition are discussed. The benefits of these technologies are also discussed.

Chapter 9, "Effect of Biofuels on Greenhouse Gases," provides comprehensive information pertaining to the main greenhouse gases in the atmosphere and in internal combustion engines, the degree of carbon-neutral fuel, and our case studies of carbon dioxide, methane, and nitrous oxide emissions from internal combustion engines, carbon dioxide emission reduction using energy efficiency improvement, carbon content reduction in fuel, and the feasibility of the carbon capture and storage system. The effects of natural gas in compression-ignition engines on greenhouse gas emissions are also discussed. Solved numerical problems related to greenhouse gases are given in this chapter.

Chapter 10, "Answers to Frequently Asked Questions," clarifies some misconceptions about biofuels and their utilization in internal combustion engines. This chapter connects the important points pertaining to biofuels made throughout the book and I hope that this information is useful as a resource.

I have written this book with careful processing of the information available in the literature and with the intent of presenting information clearly in an easy-to-understand format. Information not well established by researchers is avoided. I welcome your feedback and suggestions and I will try to incorporate them in a revised edition. I am looking forward to your positive reply.

The book is aimed at dissemination of information pertaining to biofuel and its utilization in internal combustion engines to many stakeholders. In addition, this book encourages researchers to strengthen biofuel programs in order to enhance sustainable energy, the environment, and ecosystems. If even a small percentage of this book's information will call readers to action, its mission will be completed. The information provided in this book will (1) allow readers to update their basic knowledge in the area of biofuels, its utilization in internal combustion engines, and its impact on the environment and ecology, (2) serve as a comprehensive reference source in hand for undergraduate, postgraduate, and PhD projects and research work, (3) be a valuable resource for researchers at universities and in industry, (4) serve as a reliable guide for environmentalists, ecologists, policy makers, and personnel and researchers from the automotive and oil industries.

Acknowledgments

I would like to thank Dr. Gagandeep Singh, Editorial Manager, Taylor & Francis, and Ms. Mouli Sharma, Editorial Assistant, Taylor & Francis, for their kind cooperation and valuable suggestions for finishing the book in a reasonable amount of time.

I thank all my doctoral students, master's students, and the executives and researchers from the R&D departments of the oil and automotive industries for their overwhelming support in finishing the book. I thank Mr. Vipin Dhyani (PhD scholar) for his enthusiastic involvement in this book by compilation of data, formatting, literature survey, citing the references in the text, and incorporating equations. I thank Mr. Anilkumar Ramesh Shere for drawing the figures and helping to format the book. I convey my thanks to Mrs. Shweta Tripathi (PhD scholar, assistant professor, IGNOU), Mr. Ashok Kumar (PhD scholar), Mr. Ramesh Jeeragal (PhD scholar), Mr. Devendra Singh (scientist E1, IIP Dehradun), Mrs. Nidhi (M.Tech. scholar), and Mr. Swapnil Ashok Ghate (M.Tech. scholar) for their active participation in terms of literature surveys, data compilation, and solving numerical problems. I convey my special thanks to Mr. S.K. Sinha (Director, DST, New Delhi) for his kind support.

I would like to thank Dr. Subhash Lahane, Dr. Venkateshwarlu Chintala, Dr. B.L. Salvi, Dr. Sunmeet Singh, Ms. Daizy Rajput (M.Tech. scholar), Mr. Vishal Bhatnagar (M.Tech. scholar), Mr. R. Charan (M.Tech. scholar), Mr. Sandeep Agrawal (M.Tech. scholar), and Mr. Tanmaya Singh (M.Tech. scholar) for compiling the literature and providing research data. I acknowledge Dr. Gopalakrishna Acharya (general manager, IOCL R&D Centre), Dr. Reji Mathai (deputy general manager, IOCL R&D Centre), Mr. Sakthivel (manager, IOCL R&D Centre), Mr. Sauhards Singh (senior manager, IOCL R&D Centre), and Dr. M. Subramanian (deputy general manager, IOCL R&D Centre) for their kind support.

I thank the director (IIT Delhi), Prof. Viresh Dutta (Head, CES), and my colleagues (faculty members from CES, IIT Delhi) for their kind support. I thank Prof. A. Ramesh (professor, IIT Madras), Prof. M.K.G. Babu (retired professor, IIT Delhi), Prof. L.M. Das (professor, IIT Delhi), Prof. B. Nagalingam (retired professor, IIT Madras), Dr. R.K. Malhotra (director general, Petroleum Federation of India), the late Dr. Sathish Kumar Singhal (retired scientist G, IIP, Dehradun), Dr. M.O. Garg (former director, IIP, Dehradun), and Mr. Sudhir Singal (former director, IIP, Dehradun) for their mentorship in my career growth. Last but not the least, I thank the kindhearted people who helped me directly and indirectly in writing this book.

I thank my wife, Magudeswari Subramanian, and my son, Akhil Subramanian, for their patience, kind cooperation, and moral support during the writing of this book.

With thanks and best wishes.

Dr. K.A. Subramanian
IIT Delhi, New Delhi

Author

Dr. K.A. Subramanian is currently working as an associate professor at the Indian Institute of Technology Delhi and he is also a former scientist. His research areas include internal combustion engines and alternative fuels. He earned his doctoral degree from the Indian Institute of Technology Madras. He was nominated for participating in a project study mission on energy efficiency, sponsored by the Asian Productivity Organization, Japan, in 2009. As a program coordinator, he has conducted several short-term quality improvement program courses on a national level for faculty members from other universities and engineering institutions. He is a member of many professional societies, including the National Productivity Organization, the Society of Automotive Engineers, and the Combustion Institute (Indian Section).

He is presently an M.Tech. coordinator, CES, IIT Delhi, for two postgraduate programs: M.Tech. in energy studies and M.Tech. in energy and environment management. He has developed some master-level courses, including those on hydrogen energy, emission control in internal combustion engines, and zero-emission vehicles. He also teaches several courses, including those on alternative fuels for transportation, hydrogen energy, energy, ecology and environment, heat transfer, power plant engineering, cogeneration and energy efficiency, fuel technology, and energy audits.

He has completed several sponsored research projects in the area of alternative fuel use in internal combustion engines/vehicles. He has recently developed a technology for the utilization of hydrogen in spark-ignition engines (with reduced oxides of nitrogen emission and backfire) for electrical power generation with coloration of premium original engine manufacture, for an oil industry's R&D Centre, and for the Ministry of New and Renewable Energy. He has also developed a technology and filed a patent in the area of utilization of biodiesel-compressed natural gas in a compression-ignition engine in a dual-fuel mode. He has been involved in other research projects, including those for biogas vehicles, alcohol-fueled engines, and energy efficiency improvement using oxy-combustion. He has both individually and jointly guided eight doctoral scholars and is currently guiding 10 doctoral scholars, and has also guided 45 M.Tech. scholars in their research project work. He has published approximately 100 research papers in reputed international journals, symposia, conferences, and workshops. He is the coauthor of the book *Alternative Transportation Fuels: Utilization in Combustion Engines* published by CRC Press (Taylor & Francis Group, 2013).

List of Symbols and Abbreviations

c_p	specific heat capacity at constant pressure
c_v	specific heat capacity at constant volume
D	engine cylinder bore diameter
Ea	activation energy
h_c	convective heat transfer coefficient
L	stroke length
N	revolution
r	compression ratio
r	rejection
Re	Reynolds number
Rs	swirl ratio
Rt	tumble ratio
Ru	universal gas constant
rw	work ratio
$\bar{S}$	piston mean speed
S	piston speed
s	specific entropy
T	temperature
V	volume
Vc	clearance volume
Vs	swept volume
v_{sq}	squish velocity
ν	specific volume
We	Weber number
α	pressure ratio
η	efficiency
ϕ	equivalence ratio
ρ	density
γ	adiabatic heat capacity ratio
τid	ignition delay

Subscripts

a	activation
ad	adiabatic
b	burned
bd	breakdown
C	clearance
c	cutoff
comb	combustion
e	exhaust

f	flame
f	formation
f	fuel
G	generation
id	ignition delay
l	laminar
L	loss
Opt	optimum
p	pressure
ref	reference
s	swept
t	turbulence
th	thermal
u	unburned
v	volume

Abbreviations

ABE	acetone–butanol–ethanol
ADP	adenosine diphosphate
AFR	air-fuel ratio
API	American Petroleum Institute
ASTM	American Society for Testing and Materials
ATDC	after top dead center
ATF	aviation turbine fuel
ATP	adenosine triphosphate
BDC	bottom dead center
BEAMR	bioelectrochemically assisted microbial reactor
BMEP	brake mean effective pressure
BOD	biological oxygen demand
BP	brake power
BSFC	brake-specific fuel consumption
BTDC	before top dead center
BTE	brake thermal efficiency
BTL	biomass to liquid
CA	crank angle
CAA	Clean Air Act
CAI	control of autoignition
CCS	carbon capture storage
CCU	carbon capture unit
CF	correction factor
CFPP	cold-filter plugging point
CFR	Cooperative Fuel Research
CHP	combined heat power
CI	compression ignition
CIDI	compression ignition direct injection

CNG	compressed natural gas
CPPP	cold plug plugging point
CRDI	common-rail direct injection
CTL	coal-to-liquid
CV	calorific value
d-EGR	dedicated exhaust gas recirculation
DEF	diesel exhaust fluid
DI	distillation index
DIT	dynamic injection timing
DME	dimethyl ether
DOC	diesel oxidation catalyst
DPT	diesel particulate trap
EBTE	ethyl tert butyl ether
ECU	electronic control unit
EGR	exhaust gas recirculation
EL	emission load
EPA	Environmental Protection Agency
ERC	Engine Research Center
EU	European Union
Fd	ferredoxin
FDM	finite difference method
FEM	finite element method
FFA	free fatty acids
FFV	flexible fuel vehicles
FID	flame ionization
FP	friction power
F-T	Fischer-Tropsch
FVM	finite volume method
GDP	gross domestic product
GT	gas turbine
GWP	global warming potential
H	enthalpy
HR	humidity ratio
HCCI	homogenous charge compression ignition
HFO	heavy fuel oil
HFRR	high-frequency reciprocating rig
HMF	hydroxymethylfurfural
HRR	heat release rate
HVO	hydrogenated vegetable oil
IARC	International Agency for Research on Cancer
IC	internal combustion
IDC	Indian driving cycle
IOs	inorganic
IP	indicated power
IPCC	Intergovernmental Panel on Climate Change
IQT	ignition quality tester
LH	liquefied hydrogen
LHV	low heating value
LNG	liquefied natural gas

LNT   lean NOx trap
LPG   liquefied petroleum gas
LTC   low-temperature combustion
LTFT   low-temperature flow test
MBT   maximum brake torque
MEC   microbial electrolytic cells
MeOH   methanol
MEP   mean effective pressure
MFC   microbial fuel cell
MIE   minimum ignition energy
MON   motor octane number
MPa   megapascal
MTBE   methyl-tertiary-butyl-ether
NAPCC   National Action Plan on Climate Change
NDIR   nondispersive infrared analyzer
NHTSA   National Highway Traffic Safety Administration
NMHC   nonmethane hydrocarbon
NS   number of strokes
NTP   normal temperature and pressure
OME   original engine manufacturer
Os   organic
PAH   polycyclic aromatic hydrocarbon
PCCI   premixed charge compression ignition
PHA   polyhydroxyalkanoates
PKOME   palm kernel oil methyl ester
PLA   polylactates
PM   particulate matter
PNG   piped natural gas
PNOME   peanut oil methyl ester
POME   palm oil methyl ester
PP   pour point
ppb   parts per billion
ppm   parts per million
PSA   pressure swing adsorption
PTFE   polytertrafloroethylene
RCCI   reactivity controlled compression ignition
RED   Renewable Energy Directive
RFG   reformulated gasoline
ROME   rapeseed oil methyl ester
RON   research octane number
RPR   rate of pressure rise
RVP   Reid vapor pressure
SCR   selective catalyst reduction
SFOME   sunflower oil methyl ester
SI   spark ignition
SMD   Sauter mean diameter
SOME   soybean oil methyl ester
SR   swirl ratio
ST   steam turbine

STP	standard temperature and pressure
TDC	top dead center
THC	total hydrocarbon
TPE	total premixed charge energy
UHC	unburned hydrocarbon
ULEV	ultra-low emission vehicle
USDA	United States Department of Agriculture
VOC	volatile organic compounds
WFOME	waste fried oil methyl ester
WHO	World Health Organization
WSD	wear scar diameter

Symbols: Reactants/Products/Chemicals

C_4	butanol
C_5	pentanol
C_6	hexanol
C_7	heptanol
C_8	octanol
C_{12}	dodecanol
C_{20}	phytol
$C_{12}H_{24}O_{11}$	sucrose
C_2H_5OH	ethanol
$C_6H_{10}O_5$	starch
$C_6H_{12}O_6$	glucose
CH_3CH_3	ethane
CH_3COCH_3	acetone
CH_3COOH	acetic acid
CH_4	methane
CO	carbon monoxide
Co	cobalt
CO_2	carbon dioxide
CuO	cupric oxide
H_2	hydrogen
$H_2C=CH_2$	ethene
H_2O	water
H_2S	hydrogen sulfide
HC	hydrocarbon
KOH	potassium hydroxide
Mo	molybdenum
$NaOH$	sodium hydroxide
Ni	nickel
N_2O	nitrous oxide
NOx	oxides of nitrogen
O_2	oxygen
PM	particulate matter
SOx	oxides of sulfur

1

Introduction to Biofuels

1.1 Overview of Fossil Fuels

The sun is the supreme source of energy for all planets including earth. Earth receives the sun energy in the form of electromagnetic radiation. A producer converts sun energy into bioenergy by the photosynthesis process, which is a biological process transforming sunlight with carbon dioxide taken from the atmosphere and water in plant cells into energy stored in the plant cell. A plant converts solar energy into biomass with carbon dioxide and water, as shown in Equation 1.1. Subsequently, the energy flows from plant to herbivore, carnivore, and scavenger, as shown in Figure 1.1. The livestock except plant are in dynamic motion.

$$6\,CO_2 + 6\,H_2O\,--\!\rightarrow\ \ C_6H_{12}O_6 + 6\,O_2 \tag{1.1}$$

The plant, herbivore, and carnivore eventually die due to aging or other causes including disease, forest fires, and natural disasters (cyclones, floods, tsunamis, and earthquakes). Then, their remains start to decay and become buried under the earth. The remains are composed of carbon and hydrogen, and the process of decay and decomposing proceeds through several decades, resulting in changes to the molecular structure. Then, after centuries, the structure converts from larger molecules into smaller molecules that have energy potential called fossil fuels, which include coal, oil, and natural gas. Thus, fossil fuels are generally formed under the earth. The carbon in the organism is recycled from one to another, as shown in Figure 1.2.

The formation mechanism of coal, petroleum oil, and natural gas is explained as follows.

Coal is formed under the earth mostly in terrestrial regions as it is mostly being explored in the coal mining located in lithosphere. The total proved coal reserve at the end of 2015 with reasonable certainty was about 891.53 billion tonnes. It may be noted that the reserve-to-production ratio of coal in the world in a typical business scenario is about 114 and this indicates that the world's coal reserve would disappear at the beginning of the next century (BP Energy Outlook, 2015). A typical piece of coal has an amorphous molecular structure consisting of carbon, hydrogen, and inorganic compounds including sulfur. Coal is used exclusively as the fuel in steam turbine power plants for electricity generation. In addition, coal is used as fuel for heating sources in areas such as the cement industries and chemical processes. However, coal as a fuel cannot be used in internal combustion (IC) engines because ash formation (due to an inorganic element/compound embedded in coal) during combustion inhibits its use in the engines. Synthetic or reformed coal as internal combustion engine fuel has received more attention in recent years than other

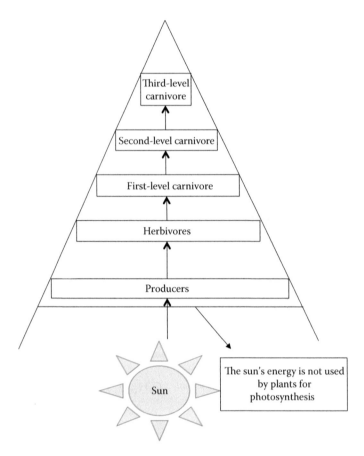

FIGURE 1.1
Energy flow (pyramid) from the sun to all organisms.

synthetic fuels. Coal can be reformed into synthetic fuels such as methanol and dimethyl ether (DME) through gasification and synthesis processes called coal to liquid (CTL) fuel. For countries such as India and China, which have high reserves of coal, CTL fuels (coal reformed fuel) can be a viable option to reduce import of their crude oil.

Petroleum Oil: Petroleum oil is formed in both terrestrial and aquatic systems. The total proved coal reserve at the end of 2015 with reasonable certainty was about 114 billion tonnes. The reserve-to-production ratio of oil in the world in a typical business scenario is about 50.7 and oil reserves would disappear in the middle of the current century (BP Energy Outlook, 2015). The main elements of raw crude oil include carbon, hydrogen, and sulfur. Crude oil contains different hydrocarbon compounds. As the fuel quality of raw crude is not desirable as IC engine fuel, crude oil properties are refined using atmospheric and vacuum distillation methods in oil refineries and the distilled fuel's molecular structure and chain length are reformed using a number of chemical processes, including catalytic cracking, polymerization, and isomerization, whereas sulfur-embedded crude oil is removed using a desulfurization process according to the desired fuel quality requirement. High octane number fuels, including gasoline and liquefied petroleum gas (LPG) obtained from petroleum refineries, are used as fuel in spark-ignition engines whereas high cetane number fuels, such as diesel, are used as fuel in compression-ignition engines. Fuel quality

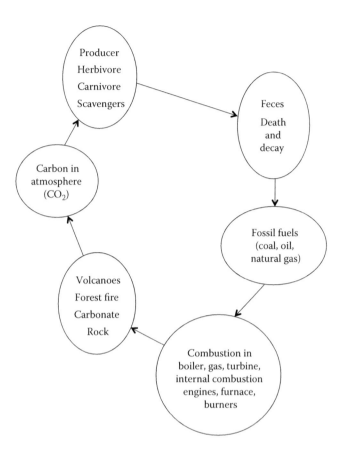

FIGURE 1.2
Carbon cycle.

is being upgraded further in order to meet the current stringent fuel quality norms. For example, the octane number of gasoline is in the range of 85 to 91, but must be increased further as advanced engines and vehicles need an octane number of 95 and above for further improvement of their fuel economy. Similarly, the sulfur content in diesel fuel has to be decreased to ultra-level (targeted level: 15 ppm). Crude oil's molecular structure is complex as it blends many hydrocarbons whose structure includes paraffin, aromatics, and olefin. The high cetane number fuel has a structure of mostly aromatics and olefin with a double bond whereas the molecular structure of cetane number fuel is mostly paraffin. The aromatic compound in petroleum fuel is a source for polycyclic aromatic hydrocarbon (PAH), which forms during combustion. However, currently PAH is not desirable and must be reduced in order to meet the current emission norms. An olefin compound is an unsaturated hydrocarbon and it is highly reactive; combustion with these compounds would lead to deposit on a heat engine's combustion chambers. Sulfur embedded with hydrocarbon is converted into oxides of sulfur (SOx) during combustion and this is one of the main emissions that contributes to acid rain and ecosystem damage. The octane number of gasoline fuel has to be increased for achieving better thermal efficiency or fuel economy of automotive vehicles. Therefore, the fuel quality of petroleum fuel has to be upgraded; however, there are lot of challenges to attain the desired property of the fuel with an affordable cost.

Fossil fuel reserves deplete at a faster rate due to large consumption, resulting in a huge gap in demand and supply. Diesel and gasoline fuels, which are derived from the feedstock of petroleum crude oil, are used as the main fuels in internal combustion engines. An internal combustion engine fueled with petroleum fuels during combustion emits harmful emissions such as carbon monoxide (CO), hydrocarbon (HC), NOx, and smoke/particulate matter. Biofuels are a viable option to tackle these issues. Fossil fuels that are generally embedded with sulfur result in the formation of SOx emissions during combustion. Sulfur molecules are generally not embedded with biofuels, which are structured with elements such as carbon, hydrogen, nitrogen, and oxygen. The composition of biofuels does not generally have aromatic content. The molecular structure of biofuels does not embed with sulfur ($C_xH_yO_z$) and aromatics but it contains oxygen, which helps to better oxidation of hydrocarbon during combustion.

Natural Gas: This gas contains mostly methane and is spread over the top surface of the petroleum well, coal mine, or natural gas hydrates. Methane (CH_4) is the smallest molecular structure (CH_4) among all hydrocarbon fuels present in natural gas. The typical natural gas composition is mainly methane, small percentages of hydrocarbon including ethane, propane, methane, propene, and hydrogen sulfide (H_2S), and traces of other inorganic substances. As natural gas has a higher octane number than that of gasoline and LPG, it is preferred for spark-ignition engines. This fuel is mostly used in gas turbine power plants for electricity power generation, as well as a heating source for many industrial applications. However, its reserves are limited.

1.2 Demand and Supply of Fossil Fuels

The world primary energy production must grow at 1.4% per annum from 2013 to 2035 to meet global energy needs. However, this energy consumption will negatively affect the increasing total carbon emissions by 25% with an increased rate of 1% per annum (BP Energy Outlook, 2015). Global vehicle fleets (commercial vehicles and passenger cars) will double from around 1.2 billion today to 2.4 billion by 2035. The consumption pattern of diesel and gasoline varies from county to country, as shown in Table 1.1 (Auto Fuel Vision and Policy 2025, 2014). It can be observed from the table that the European Union had the

TABLE 1.1

Gasoline and Diesel Consumption in Selected Regions/Countries in 2010

Serial Number	Region/Country	Gasoline (MTPA)	Diesel (MTPA)	Share of Diesel (%)
1	India	14	60	81.0
2	European Union	93	287	75.6
3	South Korea	8	19	70.4
4	China	69	147	68.0
5	Australia	14	16	54.4
6	Japan	43	41	48.9
7	Canada	32	27	45.2
8	United States	385	186	32.6

Abbreviation: MTPA, million metric tonnes per annum.

highest consumption of diesel—about 287 MTPA in 2010—whereas the United States was the highest in gasoline fuel consumption. In general, countries such as India, South Korea, China, Australia, and the European Union consumed over 50% more diesel as compared to Japan, Canada, and the United States. Transport fuel demand continues to be dominated by oil (89% in 2035) and the rest of the demand (11%) is met by alternative fuels. The consumption of diesel fuel in the world transportation sector is projected from 34.45 quadrillion British thermal units (BTUs) in 2010 to be 51.5 quadrillion BTUs in 2040 (International Energy Outlook, 2016). The percentage of increase in diesel consumption is about 50%.

The human population will increase to 8.7 billion by 2035 and energy consumption per capita will increase correspondingly. The gross domestic product (GDP) of the world must also increase to provide the basic products and services that are needed for this increased population to live a comfortable life, eliminate poverty in developing countries, and maintain desired levels of food and nutrition in developed countries (Equations 1.2 and 1.3). However, forest cover may shrink as areas of agricultural land need to be increased to produce the required food products for this growing population. The increase in human population, which as discussed above is inevitable, also results in severe damage to the environment and ecosystems as the habitat area in the forest for animals will be reduced, affecting the natural cycle process.

$$\textit{Human Population increase } \alpha \textit{ Gross Domestic Product (GDP) increase} \qquad (1.2)$$

$$\textit{GDP increase } \alpha \textit{ Energy Consumption increase} \qquad (1.3)$$

The fuel quality of fossil fuels that will need to be upgraded may result in incurring additional cost. The increase in human population and economic activities will result in huge energy consumption. However, as stated above, these energy demands cannot be met by fossil fuels as they are being depleted at a faster rate, resulting in a severe energy crisis. The exploration of alternative fuel and biofuels could be the solution to fill the supply gap. The selection of biomass feedstock for biofuel production is important for sustainable energy ecosystem development. If the biomass feedstock is obtained from agricultural land, this would conflict with food sources and this action will not yield any positive results. The world faces the challenge of an increasing population, which may result in a cascade effect of expanding agricultural land by reducing forest area. Biofuel activities should not lead to further reduction in forest cover or an increase in agricultural areas, and should not conflict with food sources. Otherwise, engaging in biofuel activities will not work. Therefore, biofuel programs need to carefully be planned and executed in order to yield fruitful results. For example, the feedstock for biofuel can be grown in degraded lands in terrestrial regions that have been declared as not suitable for agriculture for food production. Algae or aquatic species, which can grow in wastewater or ocean water, can be potential feedstock for biofuel production. If the biofuel's feedstock is grown in degraded land or wastewater, the habitats and food web would increase, which would have positive benefits for livestock and ecosystems. The carbon dioxide (CO_2) in the atmosphere will naturally be sequestrated by the biomass plant. Biofuel activities could keep the environment clean and green to maintain a sustainable ecosystem. Energy's demand-supply gap widening due to the depletion of energy sources as well as deterioration of the environment and ecosystems is a major concern. Renewable fuel or biofuel is a solution to these problems from a long-term perspective.

1.3 Role of Internal Combustion Engines for Mass and Passenger Transportation

Figure 1.3 shows the link between industries and consumers. Finished products are transported from upstream industries to downstream consumers. Trucks travel on rural roads and highways until they reach their final destinations, which include urban cities. Air pollution is generally low in rural and highway areas as the density of industries is less. However, air pollution in urban cities is usually at alarming levels because economic activities are high. In addition, the transportation vehicles themselves further intensify pollution levels. Cities need essential commodities including milk, vegetables, cement, and steel, and these goods are transported by automotive vehicles, as shown in Figure 1.4. The movement of people flows from other places to urban cities, and vice versa, for business trips, tourism, and other purposes. Therefore, urban cities are facing the problem of rising levels of air pollution. In many cities, power plants that are situated in urban enclaves have been closed or shifted to remote places in order to reduce air pollution. But cities still face the challenge of how to reduce air pollution created by automotive vehicles. Measures such as closing power plants and banning old and diesel-fueled vehicles are being taken by governments to reduce air pollution in many cities, including New Delhi. However, Delhi and other cities such as Beijing are still experiencing pollution levels at alarming rates mainly due to transportation vehicles.

The total cost of goods transportation can be calculated using Equation 1.4. The cost of goods and mass transportation depends on fuel cost. Fuel cost depends on crude oil prices, which fluctuate depending on its demand in the market. Sometimes, crude oil production decreases due to technical problems in the production system or sabotage of crude oil exploration and production due to military conflict between oil-producing countries and other countries. It is obvious that this cost increases as crude oil exploration reaches depleting levels. This cost can be controlled with biofuels because these fuels are renewable.

$$C_G = T_N \times D_T \times FE_A \times C_f \tag{1.4}$$

where
C_G = Total cost of goods transportation ($)
D_T = Total distance traveled (km)
T_N = Number of trucks = total weight G of goods/payload per truck

FE_A = Average fuel economy (liter/km) = $1/n \sum_i^n FEi$
C_f = Fuel cost ($/liter)

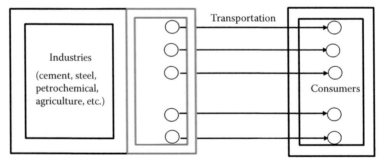

FIGURE 1.3
Transportation of goods from industries to consumers using trucks.

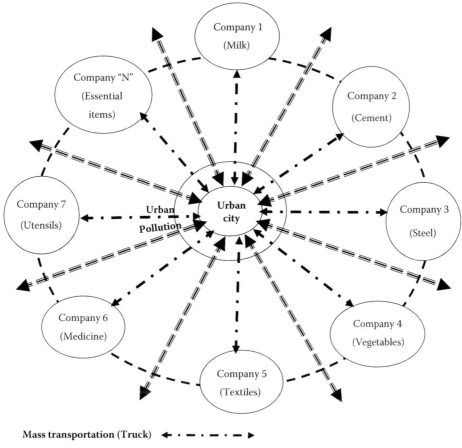

FIGURE 1.4
General scenario of transportation-associated urban air pollution.

1.3.1 Intratransportation within a Country Region

1.3.1.1 Transportation by Roadway

Finished products from manufacturers have to be distributed to end users by trucks and trains. IC-engine-based trucks are preferable for transportation of goods if rail infrastructure or connectivity is not available from the manufacturer. Note that loading (upstream: manufacturer) and unloading (downstream: end users) of the goods is a one-time activity for IC-engine-based trucks. A typical transportation flow from manufacturers to end users is as follows:

Manufacturer warehouse → Loading of goods to IC-engine truck at the manufacturer → Goods loaded and truck traveling on road → Unloading of the goods at the final destination

Examples of road transport: All heavy machines, vegetables, milk products, almost all industrial products.

1.3.1.2 Combined Mode of Mass Goods Transportation by Road and Railway

Some manufacturers have rail connectivity in which the trains go to the manufacturer to load the goods or products and then travel to the destination. Subsequently, IC-engine-based trucks are generally used for further transportation of the goods from the railway station to the end users. The drawback of this combined transportation mode is that the goods are loaded and unloaded two times: once at the manufacturer and railway station, and again at the railway station and end user. For example, a heavy machine is loaded onto the train at the manufacturer and then unloaded at the railway station. Then, the machine is loaded onto the IC-engine truck and unloaded at the final destination. This combined use of rail and roadways results in higher transportation costs due to additional labor costs and rental costs of cranes and other accessories used for loading and unloading. There can also be other issues that result from frequent loading and unloading that add to the cost, such as insurance due to increased chances of goods getting broken or damaged. The transportation cost by rail is generally cheaper than by road. If these logistics costs are higher, rail transportation may not be attractive.

Even though goods production is a continuous activity, it takes time to accumulate the finished products in bulk quantities. The main advantage of an IC-engine truck over rail transportation is that the goods—which can weigh from only a few tonnes to many tonnes—can be transported using a single truck or multiple trucks based on availability of the product in the warehouse and demand in the market. With rail, manufacturers have to wait until a minimum capacity of goods is produced because a minimum loading capacity per trip is needed for economic reasons. In addition, manufacturers prefer distribution of their products as quickly as possible to maximize profit since they are spending large amounts money for personnel, inventories including raw materials, and so forth. If the finished products are stored in warehouses for long periods, it may result in the manufacturer experiencing financial loss. Product transportation using trains generally happens a few times in a year. Railways can only do bulk transportation whereas IC-engine trucks can do both light and bulk transportation. If market demand is high for a product, it can't be supplied by rail immediately, but an IC-engine truck can deliver it quickly. The flow of combined road and rail transportation is given below:

> Manufacturer warehouse-----→ loading of goods onto trains--→ Trains travel--→ Unloading of goods from trains and loading again onto IC-engine truck in the railway station--→ The loaded truck travels----→ Unloading from truck at the final destination

Cement, steel, petrochemicals, and bulk quantity goods are examples of items that use combined transport.

1.3.2 Intertransportation: Country to Country

1.3.2.1 Combined Mode of Goods Transportation by Road and Waterways

Combined modes of goods transportation includes road and water or rail and water. For example, crude oil is transported from one country to another country through ships. This is the cheapest transportation cost compared to all other modes. However, the drawback is that it takes relatively longer to deliver the goods. This mode is only possible for countries that have a waterway infrastructure. The flow of the combined mode of goods transportation with trains and ships is given below:

Manufacturer---→ Loading to truck--> Unloading to ship--> Transportation by ship---→ Unloading to truck---→ Final destination

1.3.2.2 Combined Mode of Goods Transportation by Road and Airway

Items such as emergency medicine and important spare parts can be transported by air even though this is the costliest transportation. Air transport is preferable if, for example, a critical medicine has a short shelf life or parts to an important machine break down and manufacturer productivity will be negatively affected. In the case of machine breakdown, in order to avoid a break in productivity and possible loss of personnel and inventory, spare parts need to be procured immediately. The flow of combined road and air transportation are given below:

Manufacturer---→ Loading onto truck--> Unloading of goods and then loading onto plane in airport--> Transportation by air--→ Unloading of goods from the plane and loading onto truck→ Final destination

It can be seen from the above discussion that IC-engine based truck transportation is still the more preferable choice due to less logistical problems, including the cost of loading and unloading of goods with faster service.

1.4 Applications of Internal Combustion Engines in Decentralized Power Generation Sectors

Internal combustion engines are also used extensively in decentralized power generation sectors. Almost all type of large-scale manufacturers have internal combustion engine generator sets that require several megawatts of electricity. The required power demand is generally met by either a gas turbine power plant or a steam turbine power plant. Currently, combined cycle power plant (gas turbine + steam turbine) use has increased in order to utilize waste heat from gas turbine exhaust as well as meeting the combined requirement of electricity and heat (in the form of steam), which is called combined heat and power (CHP). Internal combustion engine generators can provide supplementary power supply to these power plants due to the following reasons:

- *Supply of Peak Load Electricity Demand.* A manufacturer's power demand will fluctuate depending on its transient electrical power requirement with respect to hours per day. The gas turbine (GT) or steam turbine (ST) power plants will generally meet the required power demand with an overload of 10% from the rated load. If the power demand is higher than the threshold load, the excess or additional power demand will be met by an internal combustion engine power plant. In certain situations—for example, lighting a space at night for a few hours—the requirement of power is less. The GT power plant will be shut down and IC engine power will be on as a heat engine's fuel economy is less at both overload and under/part load. Power plant operation during these situations is not economical. Therefore, IC engines are mandatory for supplementary power generation in large-scale manufacturing.

- *Power Backup.* If any unit of a large-scale power plant (ST or GT) fails or stalls due to unforeseen conditions, the IC engine power plant could supply the required power demand until the repaired system returns to being functional. Internal combustion engines can supply the power in continuous mode (24/7). Otherwise, stoppage of the power plant would severely affect the company's productivity, profits, and financial situation due to a large outflow of money to personnel and materials inventory. Note that the unit power cost from an IC engine generator is higher than for other power plants. However, the cost of personnel and inventory is much higher for other power plants than for an IC engine power plant. In this situation, IC engine power supply may be more economical.

- *Power Supply to Emergency Devices Including Firefighting Equipment.* If a fire occurs, fluids, including water, carbon dioxide, and foam, have to be injected into the fire spread zone in order to extinguish it. Accessories, including compressors and other devices, need electric power, which is supplied by the IC-engine power generator.

- *Quality Control.* Certain manufacturing processes, such those involving food preservation, need electricity for continuous operation because the system has to be kept at a prescribed temperature, relative humidity, and pressure, or the products would spoil. Similarly, the food processing industry also needs power in a continuous mode. If the power fails even for only few minutes, the quality of a product may deteriorate or will not meet the desired quality.

- *Pilot-Scale Plant Operation for Shorter Time Periods.* Any developed laboratory-scale product has to be studied further in a pilot scale before commercializing it in industry. The pilot-scale system needs electrical power temporarily anywhere from a few months to years. If the pilot system is successful, it will be integrated into the main system of the respective manufacturer and power supply will be taken from the main power plant. If the system fails or is unsuccessful, the technology may be abandoned. In these cases, a short duration of power demand can be met by an internal combustion engine generator set.

1.5 Off-Road Applications

IC engines are used in tractors, as the prime mover for road and building construction equipment, and in tillers, water pumps, lawn mowers, and saws. Tractors are used for agricultural work. Food prices also increase with an increase in diesel prices. There is a lot of interest in producing biofuels from biomass feedstock, which is readily available in agricultural field land. For example, wheat and rice are cultivated and produced from agricultural field land. Upgradation of the raw wheat and rice is done using farm equipment fueled with petroleum fuel, adding value to these food products. Then the rice stalk or wheat stalk can be used as feedstock for biofuel production.

1.6 Emission Load in Urban Cities

A finished product has to be transported from the manufacturer to its final destination. The transportation can be classified into four categories: road, rail, air, and water. Automotive vehicles are driven on roads and the goods are transported to their final destination through road routes. Trains run on rails that connect upstream points to downstream ones. Petroleum fuels are mostly used in automotive vehicles. When the vehicles are running, they emit harmful emissions such as CO, HC, NOx, and particulate matter (PM). Automotive vehicles, including buses, trucks, and three- and two-wheelers, are driven in inner cities. The total emission load from cities can be calculated using Equations 1.5 and 1.6.

$$\text{Total emission load from a particular class of vehicle (EL)} = \sum_{i}^{n} E_i \times D_i \times V_i \qquad (1.5)$$

$$\text{Total emission load from all class of vehicles} = \sum_{i}^{n} EL_i \qquad (1.6)$$

where
E_i = Emission factor (g/km)
D_i = Distance traveled by vehicles (km)
V_i = Total number of vehicles

Emissions controls implemented in transportation sectors could immediately help to reduce the pollution level in cities. However, as the number of vehicles sharply increases, emissions levels need to be adjusted, otherwise the current levels will not yield the desired results. For example, if CO emission is changed from 2 g/km to 1 g/km, the total load of the emission would decrease significantly. However, if the vehicle population is increased two times compared to the base reference scenario, the emission load remains the same and there is no advantage to the newer and more stringent emissions controls.

The International Agency for Research on Cancer (IARC), which is part of the World Health Organization (WHO), classified diesel engine exhaust as carcinogenic to humans (Group 1). The exhaust emissions from diesel-fueled engines increase the risk for lung cancer (WHO, 2016).

In general, air pollution levels in most urban cities are at alarming levels mainly due to transport vehicles. Biofuels as supplementary fuels could be an immediate solution to minimize the pollution problem. For example, biodiesel utilization in compression ignition vehicles could reduce particulate matter significantly. If transport fleets operated completely with biofuels, it would strengthen the sustainable energy system.

1.7 The Need for Bioenergy/Biofuels

There are several reasons why biofuels should be utilized in internal combustion engines. Some important reasons are

- Better fuel quality as compared to petroleum fuels
- Less emissions from the engine due to absence of sulfur and aromatics and presence of fuel-embedded oxygen
- Biodegradable property
- CO_2 synthesis by bioplant
- Carbon-neutral fuel
- Oxygen generation
- Habitat expansion for species
- Food web expansion
- Helps to maintain a better nutrition cycle
- Increase in soil fertility (soil quality)
- Decrease in soil erosion
- Water quality improvement by algae species

1.8 Potential of Biofuel Resources

It was previously stated that the energy received from the sun transforms to herbivores and carnivores through biomass plants. The plants grow in the biosphere, which exists in both the terrestrial system as well as the aquatic system. The earth can be divided into four major spheres: lithosphere, hydrosphere, atmosphere, and biosphere. Most livestock exists in the biosphere, where most of the organisms live together and depend on one to another. A bioplant receives energy from the sun by converting carbon dioxide and water under sunlight into glucose through the photosynthesis process. The plant uses some portion of the energy for its respiration and stores remaining of the energy in the stem, seeds, and leaves. Then, the plant expels some of the unused energy through feces and radiation. The stored energy in the plant stem is feedstock for the production of biofuels or bioenergy. A line diagram indicating first-, second-, and third-generation biofuels produced from feedstock is shown in Figure 1.5. First- and second-generation biofuels are classified as the biomass grown in terrestrial systems whereas third-generation biofuels are produced from biomass feedstock grown in aquatic systems. This figure shows the sustainability of biofuels with respect to their production from feedstock and without conflict with the food system. It can be concluded from the above-stated points that if biofuel is not produced from the proper feedstock, it cannot provide sustainable development of energy. As well, it could also give negative results if it conflicts with food production.

The leaves and stem of a plant are solid substances that contain molecular structures of carbon, hydrogen, nitrogen, and inorganics. This is called bioenergy, which can be used as boiler fuel. The seeds of the plant contain carbon, hydrogen, and volatiles. The volatiles in the seeds can be extracted and then fuel quality can be upgraded. Details and sources of bioenergy and biofuels are given in Table 1.2.

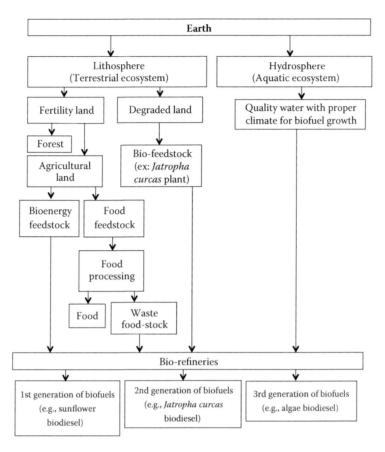

FIGURE 1.5
General routes of biofuel generation from earth's terrestrial and aquatic systems.

TABLE 1.2

Sources of Bioenergy and Biofuels

Type of Ecosystem	Organism	Energy Usage	Biofuel/Bioenergy from Unusedilized Energy	Biofuels
Terrestrial system	Plant	• Some portion of energy for respiration • Remaining portion of energy stored in stem, leaves, seeds, straw	• Energy available in seeds and leaves • Energy available from the stem after the plant dies	• Biodiesel, biogas, producer gas, ethanol, butanol, dimethyl ether, hydrogen
	Herbivore/ carnivore/ scavengers	• Respiration • Energy stored in body parts in the form of mainly fat and muscle	• Feces • Energy in the species after it dies	• Biogas • Animal-fat biodiesel
Aquatic ecosystem	Algae	• Respiration • Energy stored in the form of lipid	• Energy available in lipid • Energy after the plant dies	• Biodiesel • Biogas

1.9 Biofuels: First Generation, Second Generation, and Third Generation

The first generation of biofuels is derived from the biomass that is the source of sugar, lipid, and starch extracted from the plant. For example, sugarcane molasses is feedstock for the production of ethanol. However, ethanol can also be produced from the feedstock of sugar but it will conflict with the food source. Soybean oil is used as an edible oil in many countries but it can also be used as feedstock for the production of biodiesel. As first-generation biofuels could lead to a conflict with the food source, these are not sustainable fuels.

Second-generation biofuels are derived from cellulose, hemicellulose, and lignin in waste feedstock or as a by-product. These feedstocks include forestry wastes and agro-residue, which can grow in degraded land.

Third-generation biofuels are derived from aquatic plants. As two thirds of our earth is covered by water in oceans, this feedstock has a high potential for biofuel production. An example of this type of feedstock is algae.

1.10 Impact of Bioenergy on Ecosystems

Ecology is the study of interactions among organisms and between organisms and their environment. An ecological system is a hierarchically nested area in which all living organisms interact with the environment (atmosphere (air), hydrosphere (water) and lithosphere (land and soil)).

An ecosystem deals with a particular biological community and its interaction with abiotic components. Examples of ecosystems include river ecosystems and marine ecosystems. Ecosystems can be classified broadly into two categories: terrestrial ecosystems and aquatic ecosystems. A river ecosystem includes its biotic components (all the birds, fish, bugs, algae, and bacteria living in that river), as well as abiotic components (the rocks, sand, soil, and water in that river).

Ecological processes involve the metabolic functions of ecosystems: energy flow, elemental cycling, and the production, consumption, and decomposition of organic matter. An ecosystem also includes the following:

- Food web
- Energy/oil source
- Cellular biomass (wood)
- Nutrients recycling
- Habitat
- O_2 release
- Increase in soil fertility
- Reduction in soil erosion
- CO_2 mitigation

1.11 Impact of Bioenergy on the Environment

The impact of bioenergy on the environment includes CO_2 sequestration by the photosynthesis process, low-level emission from vehicles (due to oxygen embedded within the fuel structure, no sulfur, and aromatics). The carbon dioxide spread over in the atmosphere is naturally sequestered by producers (e.g., trees). If the producer density increases, the CO_2 in the atmosphere will decrease relatively. Plants generate oxygen during the photosynthesis process. If the CO_2 in the atmosphere decreases, the dissolving of CO_2 in ocean water will also decrease, resulting in a positive effect on marine species. Noted that the pH value of water decreases with increasing CO_2 concentration in oceans.

1.12 Comparison of Bioenergy with Other Renewable Energy Sources

Renewable energy, including solar and wind, does not contribute pollution into the atmosphere. However, these systems are unable to sequestrate CO_2 emission in the atmosphere and also cannot generate O_2. The ecosystem parameters do not change with these renewable energy systems. Among the renewable energy systems, bioenergy and biofuels have the unique feature of being able to help maintain the environment and ecology of the earth whereas the other renewable energy systems do not have any effects on the ecology of the earth.

1.13 First- and Second-Generation Biofuels

First-generation biofuels have a food-versus-fuel conflict, while second-generation biofuels are produced from waste biomass (e.g., wheat rusk) and waste water (e.g., algae biomass).

First-generation biofuels are produced from primary crops such as *Jatropha curcas* or *Pongamia pinnata*. But if farmers favor cultivating energy crops instead of food crops, this could affect food production and supply.

Second-generation biofuels are derived from the waste of biomass or agricultural products. The types of wastes include bagasse, rice straw, wheat straw and cotton stalk. Second-generation biofuels do not lead to conflicts with food production.

The major difference between a biocrop and an energy crop is that a biocrop is beneficial in all areas listed in Table 1.3, except development of a food web. Note that while other renewable energy sources (solar, wind, geothermal, etc.) do not add pollution to the atmosphere, they do not provide the widespread benefits to an ecosystem that bioenergy does. Renewable energy may also affect the habitats of organisms. For example, solar-powered systems occupy huge land areas, resulting in shrinkage of habitats due to evacuation of species to other places. Due to these factors, bioenergy stands out from all renewable energy sources because it allows sustainability of the environment and ecosystems.

TABLE 1.3

Comparison between Biocrops and Other Renewable Energy Systems in the Environment and Ecosystems

Serial Number	Description	Impact of Food Crop	Impact of Energy Crop	Other Renewable Energy Systems (Solar, Wind)
1	Food web	Good	Nil (mostly conflict with food system)	No effect
2	Energy	Moderate energy from waste	Good potential	Good potential
3	Wood	Less potential	High potential	Nil
4	Nutrient recycling	Excellent	Excellent	Nil
5	Habitat	Good but intermittent	Excellent and permanent	Negative effect as huge land areas occupied by solar-powered systems
6	O_2 release	Excellent	Excellent	Nil
7	Increase in soil fertility	Excellent	Excellent	No effect
8	Reduction in soil erosion	Excellent	Excellent	No effect
9	CO_2 mitigation	Excellent	Excellent	No effect on mitigation but does not have any emissions and will not add pollution to atmosphere

1.14 Biofuel: Carbon-Neutral Fuel

Biofuel is known as a carbon-neutral fuel. The carbon stored in a plant is the carbon dioxide in the atmosphere that is absorbed during the photosynthesis process. If the plant burns due to a forest fire or combustion with fuel wood, the carbon in the plant is then released back into the atmosphere. Therefore, at end of a plant's life cycle, the carbon stored in the plant releases an equal amount of carbon back into the atmosphere. In this way, carbon in the plant and in the atmosphere is balanced, which is called carbon-neutral because there is no new addition of carbon/carbon dioxide into the atmosphere. Biofuels that absorb and release carbon back into the atmosphere are called carbon-neutral fuels, as shown in Equation 1.7. Therefore, carbon dioxide released from a heat engine fueled with biofuels will not contribute to increasing emissions in the atmosphere.

$$C_p - C_r = 0 \qquad (1.7)$$

where
C_p = Carbon stored in the plant
C_r = Carbon released into the atmosphere

1.15 Biofuel Policy

Different countries have created different policies to adopt biofuels as fuel for transport sectors. European Union (EU) member states are required to use 10% of transport energy from renewable sources, mainly biofuels, by 2020 (EU, 2016). Biofuels and the Renewable Energy Directive (RED) formulated sustainability criteria to prevent some direct land-use change. These criteria do not address the effect of expanding agriculture land, which is called indirect land-use change (ILUC). Similarly, other countries have adopted biofuel policies as shown in Table 1.4.

TABLE 1.4

Biofuel Policy in Different Countries

Country	Biofuel	Notes	Reference
United States	Ethanol, biodiesel	• Currently all gasoline vehicles fueled with 10% ethanol-gasoline blend (E10) • Light-duty vehicles (make from 2001 and beyond) with E15 • Flexible fuel vehicles (FFVs) run with E85 • 10 and 1500 million gallons of biodiesel were consumed in 2001 and 2015 • B5 or less percentage can be used in any diesel-fueled automotive vehicles • B20 can be used in school and transit buses, mail trucks, garbage trucks, and military vehicles	U.S. Energy Information Administration (EIA)
Brazil	Ethanol, biodiesel	• Mandate of use of ethanol-gasoline blend E27 in internal combustion engines/vehicles • Vehicles run with E15, E85, and E100 • 7% Biodiesel-diesel (B7) blend to be implemented from September 2016	USDA Foreign Agricultural Service (global agricultural information index)
China	Ethanol, biodiesel	• Currently E10 is used in vehicles in some states • Planning to implement E10 in the entire country • E15 is to be implemented by 2020 • No mandate for biodiesel use in vehicles	USDA Foreign Agricultural Service (global agricultural information index)
India	Ethanol, biodiesel	• E5 is implemented in the entire country • Planning to implement E20 • Planning to implement biodiesel-diesel blend in diesel vehicles	National policy on biofuels, MNRE Government of India
Australia	Ethanol, biodiesel	• E6 and E3 are expected to implement in New South Wales and Queensland, respectively • Government is targeting to mandate E10 throughout the country in near future	Biofuels Association of Australia USDA Foreign Agricultural Service (global agricultural information index)
EU	Ethanol, biodiesel	• 5.75% Ethanol-gasoline blend is mandated for all spark-ignition vehicles • Targeting E10 by 2020 in EU member countries	European Commission
Germany	Ethanol, biodiesel	B10 is mandated in some states	Biofuels Digest
France	Ethanol, biodiesel	Targeting of 10% biofuel blend by 2020	Biofuels—The Fuel of the Future
United Kingdom	Ethanol, biodiesel	• Vehicles run with E4.5 • Planning to implement E10	Biofuels Digest

Summary

The internal combustion engine plays a pivotal role in mass and passenger transportation, as well as in decentralized power generation sectors. However, these sectors face challenges, including high transportation and unit power costs, upgradation of fuel quality, and high levels of emissions. Biofuels could be an alternative to fossil petroleum fuels and some of the challenges faced by these sectors could be mitigated with biofuels. In addition, if biofuels are produced in a sustainable manner, the biofuel program could provide tangible benefits, including overcoming the energy crisis, enhancement of fuel quality, mitigation of greenhouse gases, boosting rural economies, increasing habitat areas for herbivores and scavengers, enrichment of soil fertility, maintaining required oxygen levels in the atmosphere, and sustaining nature. Therefore, biofuels could become supplementary carbon-neutral fuels for internal combustion engines, which would assist in maintaining a sustainable environment and ecosystems.

Solved Numerical Problems

1. In an urban city, a passenger travels mostly using (1) three-wheelers, (2) four-wheelers, (3) buses, and (4) mass transportation of goods using trucks. CO emission of the particular class of vehicle as per the city's driving cycle is measured using a chassis dynamometer and the CO emission factor is calculated based on the average specific CO emission with the desired weighting factor. The CO emission factor for (1) three-wheelers, (2) four-wheelers, (3) buses, and (4) trucks is 0.02, 0.01, 0.03, and 0.04 kg/km, respectively. The average distance traveled by the vehicles is 300, 500, 1000, and 800 km per day, respectively. The vehicle population in the city is (1) 30,000, (2) 80,000, (3) 15,000, and (4) 10,000. Calculate total emission load from the fleet per day.

   Input Data

   Emission factor (EF): 0.02 kg/km (EF_1) for the three-wheeler, 0.01 kg/km (EF_2) for the four-wheeler, and 0.03 for buses and 0.04 kg/km (EF_3), for trucks

   Average distance traveled by class of vehicles (DT): 300 km (DT_1) by the three-wheeler, 500 km (DT_2) by the four-wheeler, 1000 km by buses, and 800 km (DT_3) by trucks.

   Population of particular class of vehicles (P): 30,000 (P_1) for the three-wheeler; 80,000 (P_2) for the four-wheeler, 15,000 (P_3) for buses, and 10,000 (P_3) for trucks.

   Total emission load in kg or tonnes (EL) = ?

   Solution

$$\text{Total emission load (EL)} = \sum_{i}^{n} EFi \times DTi \times Pi \qquad (1.8)$$

Total CO emission load (EL) = $0.02 \times 300 \times 30{,}000 + 0.01 \times 500 \times 80{,}000$
$$+ 0.03 \times 1000 \times 15{,}000 + 0.04 \times 800 \times 10{,}000$$
$$= 1{,}350{,}000 \text{ kg of CO/day or 1350 tonnes of CO/day}$$

2. An average NOx emission in a transportation fleet in 2014 is 4 g/km. The new emission norm for NOx of the fleet is 2 g/km and all vehicle technologies are upgraded to meet the new norms. The new emission norms for the fleet were implemented in 2015. If the population vehicle increases annually at a rate of 10%, when (what year) does NOx emissions equal the same level of 4 g/km (reference year of 2014)? Assume the distance traveled by the vehicle fleet stays the same.

Input Data

NOx emission in 2014 = 4 g/km

New NOx emission in 2015 = 2 g/km

Simple and compound annual rate of increase of vehicle fleet population = 10% (reference to 2014)

How many years and in what year is the NOx emission of the fleet equal to the reference year 2014?

Solution

$$\text{Total emission load (EL) in 2014} = \sum_i^n EF \times DT \times (P + PNR/100) \qquad (1.9)$$

$$\text{EL 2014} = 4 \times DT \times P \qquad (1.10)$$

$$\text{ELn} = 2 \times DT \times \left(P + \frac{PNR}{100} \right) \qquad (1.11)$$

If EL and DT in 2014 are equal to ELn, then Equation 1.10 is equal to Equation 1.11. If the vehicle population increases annually at the simple rate of 10%:

$$4 \times DT \times P = 2 \times DT \times (P + P \times N \times 10/100)$$

Total number of years (N) = 10 years and the emission level is the same in 2025. If the vehicle population increases compounded annually at a rate of 10%:

$$4 \times DT \times P = 2 \times DT \times P + (1 + 10/100)^N$$
$$2 = 1.1^N$$
$$N = \log 2/\log 1.1 \approx 7.3 \text{ years}$$

Total number of years (N) = 7.3 years and the emission level is same in 2022.

This solution indicates that the NOx emission load in the fleet in 2022/2025 would be equal to the reference year (2014). The new emissions norm is ineffective and does not provide the desired results if the vehicle population increases rapidly. Therefore, the emissions norm has to be upgraded periodically in order to control air pollution levels in the atmosphere.

References

Auto Fuel Vision and Policy 2025, Government of India, 2014.

Biofuels—The fuel of the future, www.biofuel.org.uk.

Biofuels Association of Australia, www.biofuelsassociation.com.au.

Biofuels Digest, Biofuels mandate around the world, 2016, www.biofuelsdigest.com.

BP Energy Outlook 2035, Feb. 2015, www.bp.com/energyoutlook.

China-Peoples Republic of, Biofuels Annual 2015, USDA Foreign Agricultural Service (global agricultural information index), www.usda.gov.

EIA, US Energy Information Administration, www.eia.gov0.

European Commission, www.ec.europa.eu.

International Energy Outlook 2016, http://www.eia.gov/forecasts/ieo/world.cfm.

National policy on biofuels, Ministry of New and Renewable Energy (MNRE), Government of India, December 2009, www.mnre.gov.in.

USDA Foreign Agricultural Service (global agricultural information index).

2

Production of Biofuels

2.1 Definition of Biomass and Biofuel

The word "bio" is derived from "bios," which originates from Greek and indicates life. Biomass is an organic matter that has stored energy through the photosynthesis process. Biofuel is a fuel derived from living matter (biomass). The sun's energy is converted into bioenergy (biomass) stored in plants (biomass) through the photosynthesis process by a biological reaction between water and carbon dioxide.

2.1.1 World Scenario of Biomass

The world's primary biomass demand for 2050 in Energy Technology Perspectives (ETP) 2008 *Blue Map Scenario* is projected at about 3606 million tonnes of oil equivalent (Mtoe), whereas the projected share of the world's transport fuel for 2050 is predicted to be about 26% (Anselm, IEA, 2010). The demand for these biofuels would be met by second-generation biofuels. The IEA reported that the feedstocks for producing second-generation biofuels are available in crop processing (coffee, rice, corn, cacao (shells, husks, cob)), sugar production industries (bagasse, pulp), vegetable oil production industries (canola, oil palm, jatropha (cake, shells, fruit bunch)), forest residue processing (sawdust, bark), municipal solid waste, and agriculture residue (straw, stover, etc.). The first-generation biofuels would lead to conflict between food verses fuel as discussed in Chapter 1, and it is a valid assumption that the second-generation biofuels would also not lead to conflict since biofuels can be produced from waste feedstock such as residue or waste (waste-to-energy concept). Almost all countries in the world are deciding to implement biofuels in the transportation and power generation sectors. Apart from the energy, a biofuel program would also have multiple benefits in terms of creating more jobs, boosting rural economies, and producing tangible products (e.g., chemicals). A biofuels program would strengthen the sustainable energy base, the environment, and ecosystems.

2.2 Raw Biomass as an Energy Source and Its Limitation as a Fuel

Biomass has energy potential because it can be used directly as fuel for cooking, heat treatment processes in industries, and furnace fuel for generating steam through boilers. However, the use of raw biomass as a fuel is being discouraged as the combustion efficiency of biomass in traditional furnaces or stoves is low due to poor combustion because improper mixing of solid fuel (heterogeneous mixture) due to less diffusivity results in incomplete combustion, and hence, high levels of carbon monoxide, unburned hydrocarbons, and particulate matter. In the case of internal combustion engines, biomass cannot be used as fuel due to poor combustion, which is due to a less homogeneous mixture (pulverised biomass particles with air) and problems of settlement of inorganic (ash) substance in the combustion chamber as the engine does not have provisions to remove ash frequently during or after combustion.

2.3 Classifications of Biomass

Biomass can be divided into five categories:

1. Biomass from wood, which is primarily derived from natural forest and woodland sources (e.g., sawdust)
2. Biomass from agricultural residues (e.g., rice husks and straw)
3. Biomass from energy crops that are grown exclusively for energy production (e.g., corn, palm, *Jatropha curcas, Pongamia pinnata*)
4. Biomass from urban waste-/refuse-derived fuel (e.g., municipal solid waste, sewage, industrial waste, household waste)
5. Biomass from aquatic ecosystems (e.g., algae)

The classification of these biomass resources is shown in Table 2.1 (Shamsul et al., 2014). The main component of biomass is cellulosic substance, which is an important component for the production of biochemicals and biofuels. Cellulosic biomass primarily contains cellulose, hemicellulose, lignin, small amounts of proteins, lipids, and ash. Cellulose is an insoluble substance and is the main constituent of plant cell walls and fibers. Lignin, which is a complex organic polymer, strengthens the plant's cell walls. Lipid (fatty acids) is also insoluble in water but soluble in organic solvents. Examples of lipids include natural oils, waxes, and steroids. Protein is a nitrogenous compound (long chains of amino acids)

TABLE 2.1

Biomass Feedstock and Their Yield

Biomass Category	Feedstock	Amount of Biomass Residue in the World (EJ/yr)
Agriculture waste	Rice husk, rice straw, wheat straw, vegetable residue, etc.	14.5
Livestock	Animal waste, butchery waste	42.5
Industry	Sewage sludge, organic	20
Household	Garbage, human waste, etc.	
Continental area (production)	Grain, plant, vegetable, fat and oil, etc.	36.5
Water area (production)	Algae, photosynthetic	
Fuel logs	Waste from fuel	14.5

Note: ExaJoule = 10^{18} Joules.

that helps to grow cells and build the bodies of all organisms. Ash is the mineral component of an organic substance that is generally the residue left after burning. Details of biomass waste for the production of biomethanol are outlined next.

2.4 Resources for Production of Biomethanol

Agricultural Waste: This is agricultural residue that is produced in fields/farms during harvesting and other agricultural activities. The residue, which is an agrowaste, includes rice husks, rice straw, wheat straw, seed cake, kernel, and other substances (stem, leaves, etc.). Lignocellulose materials from agriculture and forest are sources of hexose (C-6) and pentose (C-5) sugars and these materials can be used for the production of biofuels, chemicals, and other by-products. Various types of agricultural waste biomass such as rice bran, straw and husks from rice, sugar cane bagasse, corncobs, and nutshells can be used as raw materials for biomethanol production. Among these feedstocks, rice bran has a high potential to produce methanol fuel with the highest yield of 55% by weight, whereas methanol with rice straw and husks have a yielding efficiency of 36% and 39% by weight, respectively (Shamsul et al., 2014).

Forestry Waste: Horticultural waste biomass, including tree trunks, twigs, and leaves, is a potential source of cellulosic biomass feedstock.

Livestock and Poultry Waste: Farmers get economic benefits if they use animal manure as a biomass source to produce biofuels. Biorefineries can collect the biomass waste from the farmers for producing biofuels.

Sewage Sludge: There is an increasing amount of chemical sewage sludge produced by industrial plants and consumers. Because sludge comprises solid waste, carbon dioxide, carbon monoxide, hydrogen and methane, biomethanol can be produced from the sludge; otherwise, the gaseous substances in it are released to the atmosphere as emissions. It may be noted that methane is one of the greenhouse gases that will contribute to global warming and climate change.

2.5 Biomethanol

Methanol is a toxic, colorless, and volatile flammable liquid alcohol. Methanol is produced using synthesis gas as a raw material through a methanol synthesis process. The synthesis gas consists of carbon monoxide (CO) and hydrogen (H_2) which are produced through a gasification process from biomass feedstock. The chemical reaction for methanol formation is shown in Equations 2.1 and 2.2 (Zhen et al., 2015).

$$CO + 2H_2 \rightarrow CH_3OH \tag{2.1}$$

$$CO_2 + 3H_2 \rightarrow CH_3OH + H_2O \tag{2.2}$$

The hemicelluloses and cellulose substance will decompose at the temperature in the ranges of 470–530K and 510–620K, respectively. Lignin is in the temperatures from 550 to 770K. Methanol is one of the outputs of a photoelectrochemical cell that uses methane or biomass. The reaction of methane with water produces methanol, hydrogen, and acetic acid at 94.1°C with the doped semiconductor photocatalyst (Zhen et al., 2015).

The production process and corresponding yield are given in Table 2.2. The general plan for biomass to methanol conversion is shown in Figure 2.1.

TABLE 2.2

Kinetic Parameters of Biomethanol Production

References	Raw Materials	Conversion Process	Yield
Gullu et al., 2001	Solid waste	Pyrolysis	185 kg per ton of solid waste
Gullu et al., 2001	Hazelnut shells	Pyrolysis	7.8%
Gullu et al., 2001	Wood	Destructive distillation	6 gallons per ton of wood
Gullu et al., 2001	Biomass	Gasification	100 gallons per ton of biomass
Galindo et al., 2007	Biomass	Gasification	12.21 kton/year
Galindo et al., 2007	Biomass	Gasification	17.60 kton/year
Kumabe et al., 2008	Wood	Gasification (biomass to liquid–MeOH)	47.7%
Dong et al., 1997	(Biomass + steam + methane)	Gasification	199.2 kg from 100 kg biomass + 1.2 kg steam + 0.5 kg methane
Dong et al., 1997	Wood + natural gas	Hynol gasification	101.5 MM gallon/year
Hasegawa et al., 2010	Wood	NEDO (gasification)	510 L
Xu et al., 2006	Conditioning biosyngas + biomass char	Gasification	1.32 kg
Weimer et al., 1996	Carbon dioxide	Gasification + electrolysis	50 kJ/mol
Nakagawa et al., 2007	Biomass: agricultural waste	Gasification	38–50%
Reno et al., 2011	Wood	Gasification	18.237 t h^{-1}
Chmielnik et al., 2003	Raw gas from coal + biomass	Cogasification	132.5 thousand m^3
Sahibzada et al., 1997	CO_2/H_2	Electrolysis	0.44 mol h^{-1}g^{-1}
Sahibzada et al., 1997	$CO_2/H_2 + H_2O$	Hydrogenation	0.043 mol h^{-1}g^{-1}
Taylor et al., 2000	Methane	Photocatalytic + thermochemical	(i) No catalyst: 0.04 mL/min (ii) With hydrogen peroxide: 0.17 mL/min

Source: Shamsul, N. S., Kamarudin, S. K., Rahman, N. A. and Kofli, N. T. (2014), An overview on the production of bio-methanol as potential renewable energy, *Renewable and Sustainable Energy Reviews*, 33, 578–588. (Reprinted with permission.)

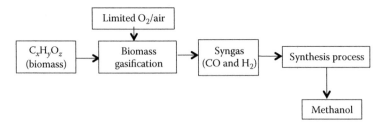

FIGURE 2.1
General plan for biomass to methanol conversion.

2.6 Ethanol

The raw materials and production process for ethanol are highlighted next.

2.6.1 Raw Materials for Ethanol Production

The main biomass for producing first-generation biofuels includes sugar beet and wheat (in Europe), corn (in United States), sugarcane (in Brazil), and sugarcane molasses (in India). The production process for producing biofuels is mainly through fermentation by converting sugar into ethanol. The International Energy Agency (IEA, 2010) states that "first generation biofuels are biofuels which are on the market in considerable amount today."

Second-Generation Biofuels: The International Energy Agency (IEA, 2010) states that "second-generation biofuels are those biofuels produced from cellulose, hemicelluloses, or lignin." The main feedstocks are wood, pulp fibers, bagasse, agricultural residues, and industrial biowastes. The main production processes are hydrolysis, fermentation, and thermochemical conversion. Cellulosic type feedstock needs a specialized pretreatment, such as an acidic process, followed by hydrolysis and fermentation processes. For example, solid waste such as bagasse is converted to synthesis gas (a mixture of carbon monoxide and hydrogen) through the gasification process. Then the syngas as feedstock is converted to bioethanol using a suitable chemical synthesis process.

Third-Generation Biofuels: The main raw materials in third-generation biofuels are algae and other biomass from aquatic ecosystems. Algae solid waste can be converted into syngas and then bioethanol. Lipids in algae is generally used for biodiesel production.

The type of production process/technology will be decided based on the availability of biomass material and well-to-tank efficiency (life-cycle efficiency). The well-to-tank efficiency indicates how much energy input is needed for ethanol production, storage, and transportation and dispensing. The energy input requirement for the fermentation process for conversion of sugarcane or starch material to ethanol is less than the energy input of the available established technology for cellulose-type biomass feedstock. The process time is shorter for starch or sugar-based feedstock than for cellulosic feedstock. Note that fermentation technology is an almost matured process and is readily available, but the technology for conversion of cellulose and other types of biomass to bioethanol is still in progress. Most ethanol is produced from starch-based crops by dry- or wet-mill processing, as shown in Figure 2.2.

2.6.2 Fermentation

Ethanol is produced by the fermentation of the sugars in molasses. Sucrose is converted into glucose or fructose through a hydrolysis reaction, as shown in Equation 2.3.

$$C_{12}H_{22}O_{11} + H_2O \rightarrow 2C_6H_{12}O_6 \tag{2.3}$$

Ethanol is then produced by the fermentation of these monosaccharide sugars (Equation 2.4). The reaction is exothermic with heat of 1200 kJ/kg of ethanol.

$$C_6H_{12}O_6 \rightarrow 2C_2H_5OH + 2CO_2 \tag{2.4}$$

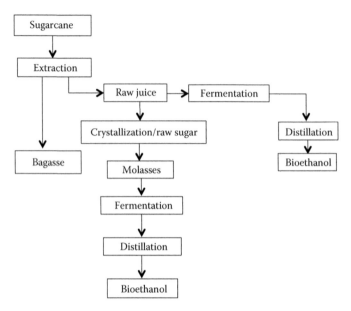

FIGURE 2.2
Ethanol production from sugar-based substances.

2.6.3 Cellulosic-Based Ethanol Production

Production of ethanol from cellulosic feedstock such as grass, wood, and crop residues is more complex than using starch-based feedstock and the process is shown in Figure 2.3. Ethanol can be produced primarily using two methods out of many methods, such as biochemical and thermo-chemical. The biochemical process needs a pretreatment to release hemicellulose sugars and hydrolysis process will break cellulose into sugars. Sugars can be fermented into ethanol and the recovered lignin is used to produce energy. In the thermochemical conversion process, solid biomass feedstock is converted to synthesis gas (mixture of carbon monoxide and hydrogen) and then the syngas is converted to ethanol using a suitable synthesis process.

Ethyl tertiary-butyl ether (ETBE) is produced from ethanol and isobutylene in a catalytic reaction. If isobutylene is produced from fossil sources (natural gas), ETBE is not completely a biofuel. The European Union uses about 22% ETBE in E10 gasoline as it does

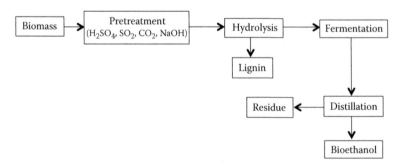

FIGURE 2.3
Ethanol production from cellulose-based substances.

not require major modification in the fuel distribution infrastructure (ETBE, European biofuels). The other ethers are methyl tertiary-butyl ether (MTBE), ETBE, tertiary-amyl methyl ether (TAME), and tertiary-amyl ethyl ether (TAEE). It is reported that MTBE could contaminate groundwater and hence be risk of cancer (MTBE, American Cancer Society). However, these issues are beyond the scope of this book as our focus is toward the discussion about more suitable biofuels for internal combustion engines.

2.7 Butanol

Biobutanol is produced through the fermentation of biomass by an acetone–butanol–ethanol (ABE) process. The fermentation is based on microorganism activity and bacteria from genus Clostridium are normally used. The common feedstocks used for the production of biobutanol are sugar and starch. The other biomasses that can be used for the production of biobutanol are barley, straw, bagasse and corn core. Figure 2.4 shows the production process of biobutanol with different feedstocks (da Silva Trindade et al., 2017):

- Biomass contains lignocellulosic, which is pretreated for using as a substrate for fermentation. Pretreatment can be done using sulfuric acid, steam, peroxide, and hydrothermal depending on the type of feedstock.
- The detoxification process is a process to remove inhibitors. This process can be done by activated charcoal, electrodialysis, overliming, and others depending on the feedstock and pretreatment method used.

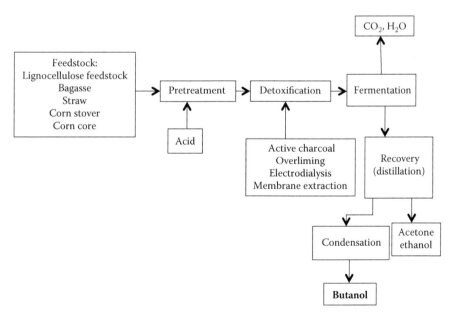

FIGURE 2.4
Schematic of the biobutanol production process.

- Fermentation is a process and used for sugar- and starch-based feedstock. Different processes can be used depending on the feedstock and pretreatment method used.
- After fermentation, purification is done and then separation of acetone, ethanol, and butanol based on their boiling points, which are 56°C, 78°C, and 118°C, respectively (da Silva Trindade et al., 2017).

Biobutanol production plants operate on a semicontinuous basis in which each fermentation takes 21 days to complete. The final product is then purified and recovered to get acetone, butanol, and ethanol in a proportion of 3:6:1 by mass (Ni et al., 2009). Equations 2.5 and 2.6 show the stoichiometric equations of ABE fermentation process with cornstarch.

$$[C_6H_{10}O_5]_n + n\,H_2O \rightarrow n\,C_6H_{12}O_6$$
$$\text{(Starch)} \qquad\qquad \text{(Glucose)}$$

(2.5)

$$12\,C_6H_{12}O_6 \rightarrow 6\,CH_3CH_2CH_2CH_2OH + 4\,CH_3COCH_3 + 2\,CH_3CH_2OH + 18\,H_2 + 28\,CO_2 + 2\,H_2O$$
$$\text{(Glucose)} \qquad \text{(n-Butanol)} \qquad \text{(Acetone)} \qquad \text{(Ethanol)}$$

(2.6)

2.8 Higher-Alcohol Fuels

Higher alcohols have four or more carbon atoms such as butanol (C_4), pentanol (C_5), hexanol (C_6), heptanol (C_7), octanol (C_8), dodecanol (C_{12}), and phytol (C_{20}). The main disadvantages of the ethanol and methanol fuels discussed above are high corrosion and increased safety issues (low flash point). Higher alcohols are less corrosive than ethanol due to their less hygroscopic nature and the fuels do not lead to corrosion problems in fuel systems (handling, storage, transportation, and dispensing) (Rasskazchikova, 2014). These fuels have high flash points as the flash points of diesel, methanol, ethanol, isobutanol, n-pentanol, n-hexanol and n-octanol are >55, 11–12, 17, 28, 49, 63, and 81, respectively (Kumar, 2016). The flash point shows that as the carbon number in alcohol fuel increases, the flash point increases and this is an advantage in terms of safety. However, the research octane number (RON) and motor octane number (MON) decrease with increasing carbon numbers in straight-chain (paraffin type) higher alcohol. The RON and MON of fuels are 109 and 89 with methanol, 109 and 90 with ethanol, 104 and 89 with n-Propanol, 98 and 85 with n-Butanol, 80 and 74 with n-Pentanol, 56 and 45 with n-Hexanol, and 28 and 27 with n-Octonol, respectively (Sarathy, 2014). As the RON is much less with higher alcohol fuels than base gasoline ethanol and methanol, the fuels may not be suitable for spark-ignition engines. Even though higher-alcohol fuels contain a paraffin type of molecular structure (straight-chain), these fuels' cetane number of methanol, ethanol, isobutanol, n-Pentanol, n-Hexanol, and n-Octanol are 5, 8, <15, 20, 23, and 37, respectively, but these numbers are much less compared to base diesel fuel (52) (Kumar, 2016). Iso-structured higher alcohol may have a higher octane number but it may need an additional process that has to be well established. Therefore, these higher-alcohol fuels are currently less suitable for spark-ignition engines and these fuels also have less scope for increasing the compression ratio of the engines for improving thermal efficiency. Therefore, these higher-alcohol fuels are not discussed further.

2.9 Biomethane and Biogas

Biomethane or biogas is produced from the feedstock of biomass by a biological process commonly known as anaerobic digestion in the temperature range of 10°C to 100°C and the moisture content is in the range of 50% to 99%. The anaerobic process is a biological process that occurs without oxygen. The anaerobic bacteria convert biodegradable matter in the feedstock into methane and carbon dioxide (Jingura et al., 2009). There are three stages in which the biological reaction takes place. In the first stage, this process is called hydrolysis, in which fermentative bacteria hydrolyze organic matter into soluble molecules. In the second stage, the processes (acidogenesis and acetogenesis) convert the molecules into organic acids. In the third or final stage, methanogenesis is a process in which methanogenic bacteria break down the acids, leading to the formation of methane and carbon dioxide. The chemical reactions for acetogenesis and methanogenesis are given in Equations 2.7 and 2.8 (Themelis et al., 2007). The chemical reaction for the overall process can be described using Equation 2.9, where an organic material (glucose) is converted into methane and carbon dioxide by the anaerobic bacteria.

Acetogenesis:

$$C_6H_{12}O_6 \rightarrow 2C_2H_5OH + 2CO_2 \tag{2.7}$$

Methanogenesis:

$$CH_3COOH \rightarrow CH_4 + CO_2 \tag{2.8}$$

Overall chemical reaction:

$$C_6H_{12}O_6 \rightarrow 3CO_2 + 3CH_4 \tag{2.9}$$

This mixture of methane and carbon dioxide is called biogas. Biogas usually contains 60%–70% methane and 30%–40% carbon dioxide depending on the feedstock used. Some traces of hydrogen, nitrogen, oxygen, carbon monoxide, hydrogen sulfide, and ammonia are also present in the biogas (Jingura et al., 2009). Feedstock such as nonedible seed cake, food waste, agro waste, poultry waste, animal manure, municipal waste, and sewage sludge can be used to produce biogas. The yield of produced biogas depends on the feedstock used. Table 2.3 shows the maximum yield of biogas with various feedstocks. It can be concluded based on the information from Table 2.3 that nonedible seed cake has a higher potential than cattle dung to produce a higher yield of biogas (Subramanian et al., 2013).

TABLE 2.3

Biogas Yield and Methane Content from Different Feedstocks

Feedstock	Gas Yield (m³/kg)	Methane (%)
Cattle dung	0.297	55
Municipal solid waste	0.308	60
Straw	0.341	51
Pig	0.4	65
Poultry	0.45	70
Grass silage	0.576	52
Jatropha curcas seedcake	0.64	66.5
Pongamia pinnata seedcake	0.738	62.5

Source: Subramanian, K. A., Mathad Vinaya, C., Vijay, V. K. and Subbarao, P. M. V. (2013), Comparative evaluation of emission and fuel economy of an automotive spark ignition vehicle fuelled with methane enriched biogas and CNG using chassis dynamometer, *Applied Energy*, 105, 17–29. (Reprinted with permission.)

2.10 Biohydrogen

Hydrogen is mainly produced from feedstock, including fossil fuels, biomass, and water. Steam reforming, coal gasification, and electrolysis of water are the main industrial methods of hydrogen production. However, both the thermochemical and electrochemical processes of hydrogen production are energy-intensive and are not always environmentally friendly. In contrast, the biological hydrogen production processes are less energy-intensive because they operate at ambient temperatures and pressures and these biological routes are environmentally friendly. In addition, these processes can use various waste materials, which facilitate waste recycling (Das et al., 2001).

The biological methods for hydrogen production are as follows:

- Direct biophotolysis
- Indirect biophotolysis
- Biological water-gas conversion
- Photofermentation
- Dark fermentation
- Microbial electrolytic cell
- Multistage integrated method

2.10.1 Direct Biophotolysis

In the direct biophotolysis method, microalgae convert solar energy into chemical energy, which is then used for splitting molecules of water during which hydrogen is released. Equation 2.10 shows the biophotolysis reaction (Ogbonna et al., 1999; Wakayama et al., 2002).

$$2H_2O \longrightarrow 2H_2 + O_2 \text{ (in presence of solar energy)} \tag{2.10}$$

Microorganisms like *Chlamydomonas reinhardtii* and *Anabaena* sp. and microalgae like green algae or blue-green algae cyanobacteria are involved in the production of oxygen and hydrogen molecules. These all work in an environment where ferredoxin, reduced ferredoxin (Fd), and reverse hydrogenase are present. Hydrogenase can work in an oxygen-deficient environment. The efficiency of the process is about 5%.

2.10.2 Indirect Biophotolysis

Indirect biophotolysis involves four steps: (1) production of biomass by photosynthesis, (2) concentration of the biomass, (3) aerobic dark fermentation (where release of 4 mol of H_2/mole of glucose and 2 moles of acetate takes place inside the cell of algae), and (4) conversion of 2 moles of acetate into hydrogen. Cyanobacteria carries out the process according to Equations 2.11 and 2.12 (Das et al., 2008; Balat et al., 2010).

$$6CO_2 + 6H_2O \longrightarrow C_6H_{12}O_6 + 6O_2 \tag{2.11}$$

$$C_6H_{12}O_6 + 6H_2O \longrightarrow 12H_2 + 6CO_2 \tag{2.12}$$

Cyanobacteria require air, water, minerals, and light. In this process, hydrogen is produced by both hydrogenase and nitrogenase. Cyanobacteria like *Anabaena cylindrica*, and *Anabaena variabilis* are involved in the process of hydrogen production.

2.10.3 Biological Water-Gas Conversion

Gram-negative bacteria such as *Rhodospirillum rubrum* and *Rubrivivax gelatinosus* and Gram-positive bacteria such as *Carboxydothermus hydrogenoformans* are involved in the biological water-gas conversion process (Melis et al., 2006; Melnicki et al., 2008). The process takes place under anaerobic conditions, in the presence of carbon monoxide, proteins like carbon monoxide dehydrogenase, and the Fe-S protein. The oxidation of carbon monoxide releases electrons that are converted through the Fe-S protein for hydrogen production. The process proceeds at low temperatures and pressures, conditions in which the conversion to CO_2 and H_2 takes place. Bacteria like *Citrobacter* sp. and *R. gelatinosus* perform water-gas conversions.

2.10.4 Photofermentation Method

Photofermentation is a process in which photosynthetic bacteria convert organic materials and biomass into hydrogen and carbon dioxide by using solar energy. The process takes place under anaerobic conditions. The optimal temperature is about 30°C–35°C and a neutral pH of 7.0. Nonsulfur purple bacteria perform this process using simple organic acids. In the absence of nitrogen and solar energy, organic acids or biomass are converted into hydrogen as described in Equation 2.13 (Das et al., 2008).

$$CH_3COOH + 2H_2O \longrightarrow 4H_2 + 2CO_2 \tag{2.13}$$

The process of photofermentation is shown in Equation 2.14.

$$[CH_2O]_2 \xrightarrow{NADPH} \text{Ferredoxin} \longrightarrow \text{Nitrogenase} \longrightarrow H_2 \tag{2.14}$$

$$\uparrow \qquad\qquad\qquad\qquad \uparrow$$

$$\text{ATP} \qquad\qquad\qquad\qquad \text{ATP}$$

2.10.5 Dark Fermentation

Dark fermentation is done by anaerobic bacteria with sugar-rich substrate in the dark. Hydrogen is produced at low temperature in the range from 30°C to 80°C. In classical thermochemical processes, dry biomass can also be used. Efficient dark fermentation is done by microorganisms like *Enterobacter cloacae, Enterobacter aerogenes, Clostridium* sp., and *Bacillus* sp. (Argun et al., 2009; Kumar et al., 2000; Fabiano et al., 2000). Biodegradable sugars like glucose, lactose, and sucrose are a suitable substrate for this process (Hawkes et al., 2002). The pH value (5–6), dwell time, and partial pressure of the gas are the controlling parameters for hydrogen production (Fang et al., 2002).

2.10.6 Microbial Electrolytic Cells

Microbial electrolytic cells (MECs), also known as bioelectrochemically assisted microbial reactors (BEAMRs), use the process of electrochemical hydrogenation for direct conversion of biodegradable substances into hydrogen (Das et al., 2008; Ditzig et al., 2007). A microbial electrolytic cell is a modified version of microbial fuel cell (MFC) and organic matter decomposes by microorganisms without air (anaerobic type) (Liu et al., 2005). During this process, electrons, protons, and carbon dioxide are formed. In a BEAMR, hydrogen is released on the cathode. *Geobacter, Shewanella* sps., or *Rhodoferax ferrireducens* bacteria are involved in this process (Ditzig et al., 2007).

2.10.7 Multistage Integrated Method

The multistage process comprises two steps: dark fermentation and photofermentation. In the first step, anaerobic fermentation of the sugars or organic wastes into intermediate products of low molecular mass (organic acids) takes place. In the second step, these are converted into hydrogen by a photosynthetic bacterium (Hawkes et al., 2007). The general reactions of the process are as follows (Equations 2.15 and 2.16):

Step 1: Dark fermentation:

$$C_6H_{12}O_6 + 2H_2O \longrightarrow 2CH_3COOH + 2CO_2 + 4H_2 \qquad (2.15)$$

Step 2: Photofermentation:

$$CH_3COOH + 2H_2O \longrightarrow 2CO_2 + 4H_2 \qquad (2.16)$$

During dark fermentation, the biomass is degraded into hydrogen and wastewater. This wastewater contains organic acids that are further treated by the photofermentation process. In the first step, in the photofermentation process, the infrared component of light is used. In the second step, hydrogen from organic acids is produced in microbial electrolytic cells in the dark. The ammonia present in wastewater can hamper the second step. The wastewater therefore should be neutralized (Das et al., 2001).

2.11 Biodiesel

Biodiesels are mainly the alkyl esters produced using feedstocks such as vegetable oil, fat, and algae biomass. Biodiesel can be categorized based on the type of feedstock used for the fuel production.

2.11.1 First-Generation Biodiesel

First-generation biodiesels are produced from edible oil resources such as soybeans, palm oil, sunflower, rapeseed, and peanut (Nautiyal et al., 2014; Sharma et al., 2017).

2.11.2 Second-Generation Biodiesel

Biodiesels that are generally produced from nonedible feedstocks, such as waste vegetable oils and fats, nonfood crops, forestry residues, and biomass sources, are called second-generation biodiesels. Nonedible oils include *Jatropha curcas*, *Pongamia pinnata*, linseed, cottonseed, neem, camelina, putranjiva, tobacco, polanga, cardoon, deccan hemp, castor, jojoba, moringa, poon, koroch seed, desert date, *Eruca sativa* Gars, Sea mango, pilu, crambe, syringa, milkweed, field pennycress, stillingia, radish, Ethiopian mustard, tomato seed, kusum, *Cuphea*, camellia, paradise, terminalia, *Michelia champaca*, *Garcinia indica*, and *Zanthoxylum bungeanum*. Other feedstock includes waste frying oil, soap-stock-based oil, and animal fats (Beef tallow, pork lard) (Sharma et al., 2017).

2.11.3 Third-Generation Biodiesel

Algae is considered a third-generation biofuel. Some types of algae are *Botryococcus braunii*, Chlorella, spirulina, pond water algae, *Chrolella vulgaris*, *Crypthecodinium chonii*, and *Cylindrotheca* sp. (Sharma et al., 2017; Nautiyal et al., 2014).

Transesterification: The chemical process of transesterification converts free fatty acid (FFA) into methyl/ethyl/butyl esters with methanol/ethanol/butanol in the presence of a catalyst (Demarrias et al., 2011). In this reaction, the ester reacts with alcohol to form another ester in the presence of a catalyst. Due to their low cost with better availability of NaOH and KOH catalysts, they are largely used as catalysts for the reaction. If the FFA is higher than 5%, this process continues with two steps. In first step, vegetable oil is treated with acid catalyst (sulfuring acid) to reduce FFA and then is followed with a base catalyst. The conversion from fatty acid to alkyl ester is shown in Equation 2.17.

$$\underset{\text{(Fatty acid)}}{\text{R-COOR'}} \quad + \quad \underset{\text{(Alcohol)}}{\text{R''OH}} \quad \longleftrightarrow \quad \underset{\text{(Alkyl ester)}}{\text{R''COOR}} \quad + \quad \underset{\text{(Glycerol)}}{\text{R'OH}} \tag{2.17}$$

A straight vegetable is produced from the feedstock of seed using a mechanical expeller or a solvent extraction method. As the physicochemical properties of a straight vegetable oil have a higher density, viscosity, and lower cetane number than biodiesel, the vegetable oil cannot be used as fuel in compression ignition engines due to poor atomization and combustion. In this extraction process, solid seed cake is a tangible product. This feed-stock can be converted to bio-oil through the pyrolysis method or biogas using the aerobic digestion method. The biogas derived from seed cake is a successful route but bio-oil production using pyrolysis is an energy-intensive process. The fuel quality of bio-oil is poor as it has a higher density (1040 kg/m^3) and water content (31%) than diesel and biodiesel and does not meet the desired quality (Biradar, 2014). Therefore, the fuel quality of bio-oil needs to be upgraded, but this is a complex process because water separation from bio-oil is not possible by simply using the evaporation method and it needs to be cracked using either thermal cracking or the hydrocracking process.

2.12 Fischer Tropsch Diesel

Franz Fischer and Hans Tropsch developed the gas-to-liquid process in the 1920s and the Fischer-Tropsch (F-T) synthesis process is named after them. The production of paraffin-type hydrocarbon fuels using the F-T process is a set of chemical reactions in the presence of a catalyst.

F-T fuel has more desirable fuel qualities, such as ultralow levels of sulfur, low levels of aromatics, and a high cetane number. These fuel characteristics enable a compression-ignition engine with F-T diesel to emit less emissions. F-T fuels are generally produced by transforming a feedstock (natural gas or coal) to a synthesis gas (carbon monoxide + hydrogen) through the gasification process with limited oxygen/air. Then, synthesis gas as feedstock is converted into liquid hydrocarbons (synthetic crude) through the F-T process. The synthetic crude is treated to produce middle distillates (F-T diesel fuel) and other products (solvents). F-T fuel is a like synthesized straight-chain hydrocarbon fuel and has a higher cetane number, which is suitable for use as fuel in compression-ignition engines.

Natural gas transportation is not economical, so it is converted to liquid fuel (F-T fuel) through the F-T synthesis process. In this way, the storage and transportation of natural gas is more economical and less complex compared to gaseous fuel. Similarly, biomethane or biomass can be converted to F-T diesel through this synthesis process.

The feedstock for this process is any carbonaceous material, such as natural gas, coal, and biomass. Reactant gases entering an F-T reactor must be desulfurized to protect poisoning of the catalysts. The formation of synthesis gas from coal or biomass is called gasification wherein the feedstock is reacted with steam or oxygen (O_2). The next step in the F-T production process is the conversion of synthesis gas into paraffin structured-hydrocarbon.

GTL fuels are also produced from methane as feedstock through two different routes such as indirect methane conversion and direct methane conversion. The main processes are air separation, gas processing, syngas production, conversion from syngas to synthesis crude/oxygenate, and upgradation. In direct methane conversion method, the feedstock (methane) is converted into syngas through steam reforming, partial oxidation, and auto-thermal reforming methods. Then, the syngas is converted to GTL fuels (methanol, F-T diesel, dimethyl ether). In direct methane conversion process, the methane, which takes chemical reaction with the reactants (carbon dioxide or water), is directly converted to GTL fuels.

The reforming method is used in commercial applications. Syngas (CO and H_2) is produced with the reactants of methane and water/carbon dioxide, as shown in Equations 2.18 and 2.19 (Lercher et al., 1999).

$$CH_4 + H_2O \rightarrow CO + 3H_2; \Delta H^0_{298\,K} = 205.92\,\frac{kJ}{mol} \tag{2.18}$$

$$CH_4 + CO_2 \rightarrow 2CO + 2H_2; \Delta H^0_{298\,K} = 247.32\,\frac{kJ}{mol} \tag{2.19}$$

The partial oxidation method is an exothermic process, where energy consumption is low, as shown in Equation 2.20 (Keshav et al., 2007)

$$CH_4 + 1/2O_2 \rightarrow CO + 2H_2; \Delta H^0_{298\,K} = -35.25\,\frac{kJ}{mol}; \tag{2.20}$$

The reactants (H_2 and CO) are converted into -CH_2- alkyl radicals and water in an exothermic reaction. The -CH_2- radicals combine in an iron or cobalt catalytic reaction form synthetic olefin and/or paraffin HC (Equations 2.22 and 2.23). The selectivity (the amount of desired product obtained per unit consumed reactant) is influenced by parameters such as temperature, H_2/CO ratio in the feed gas, pressure, and the catalyst type. This set of processes is known as the gas-to-liquid method. The synthesis gas formation and its reaction to form paraffin and olefin types of product are given in Equations 2.21 to 2.23 (Gill et al., 2011). Figure 2.5 shows the layout of the F-T diesel production process.

$$2CH_n + O_2 \rightarrow nH_2 + 2CO \tag{2.21}$$

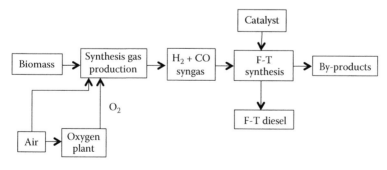

FIGURE 2.5
Fischer-Tropsch diesel fuel production flow chart.

F-T synthesis process:

$$n\text{CO} + (2n+1)\text{H}_2 \rightarrow \text{C}_n\text{H}_{2n+2} + n\text{H}_2\text{O}; \text{ paraffins} \tag{2.22}$$

$$n\text{CO} + 2n\text{H}_n \rightarrow \text{C}_n\text{H}_{2n} + n\text{H}_2\text{O}; \text{ Olefins} \tag{2.23}$$

2.13 DME

DME is produced using feedstock including methane and biomass. The molecular structure of DME is $\text{CH}_3\text{-O-CH}_3$, which indicates the absence of any C–C bonds, whose structure is significant when the fuel is combusted in a heat engine because the engine would emit less smoke or be smokeless. In addition, an important salient feature is that DME fuel is in a gaseous state in an normal temperature and pressure (NTP) ambient condition but it is in a liquid state when compressed to 5-bar pressure (by compression, critical pressure = 52.6 atm and temperature = 127°C). Therefore, combustion is better with gaseous DME and the fuel tank size requirement for its storage is less like LPG. As DME is in a gaseous state at an atmospheric condition, it mixes well with air, leading to less formation of smoke/PM emission during combustion in a compression ignition.

The C–O bond energy (358 kJ/mol) is less than that of the C–H bond (414 kJ/mol) and the C–O bond breaks easier than the C–H bond. Therefore, it can be interpreted that the fuel may need lower activation energy due to a lower self-ignition temperature or higher cetane number. Ignition delay comprises physical delay and chemical delay because physical delay relates to flammable mixture preparation whereas chemical delay refers to chemical reaction and ignition. The physical and chemical delays of DME are shorter than those of diesel, biodiesel, or F-T diesel. This is advantageous for a reduction in transient emissions and improvement of performance of compression-ignition engine with F-T diesel.

DME is produced using the following methods (Arcoumanis, 2000; Arcoumanis et al., 2008):

- Dehydrogenation of methanol
- Direct conversion from synthesis gas (syngas)

2.13.1 Dehydrogenation of Methanol

Methanol in a chemical reactor (with zeolite catalyst) at a desired temperature up to 250°C and 16.8 atm pressure transforms (Ateka et al., 2016) to DME due to the exothermic reaction by rearranging its molecular structure with the tangible product of water molecules. DME is separated by a distillation process, as shown in Equation 2.24.

$$2\text{CH}_3\text{OH} \leftrightarrow \text{CH}_3\text{OCH}_3 + \text{H}_2\text{O}; \Delta\text{H}^0 = -50.6\,\frac{\text{kJ}}{\text{mol}} \tag{2.24}$$

The direct conversion method is the simultaneous production of DME and methanol from syngas using appropriate catalysts.

First step: Conversion of the biomass feedstock to synthesis gas by gasification/steam reforming process/partial oxidation method/autothermal reforming (Semelsberger et al., 2006).

Second step: Conversion of synthesis gas to methanol through a suitable synthesis process using a suitable catalyst, such as a copper-based catalyst, as shown in Equations 2.25–2.27.

$$CO + 2H_2 \leftrightarrow CH_3OH; \Delta H^0 = -90.84 \frac{kJ}{Mol} \tag{2.25}$$

$$CO_2 + 3H_2 \leftrightarrow CH_3OH + H_2O; \Delta H^0 = -49.43 \frac{kJ}{Mol} \tag{2.26}$$

$$CO + CO_2 + 5H_2 \leftrightarrow 2CH_3OH + H_2O; \Delta H^0 = -49.43 \frac{kJ}{Mol} \tag{2.27}$$

Third step: Dehydrogenation of methanol to DME using alumina- or zeolite-based catalysts (Troy et al., 2006).

Methanol and DME can be produced from same feedstock. Methanol fuel, which has a high octane number, is suitable for spark-ignition engines whereas DME, which has a high cetane number, is suitable for compression-ignition engines.

The biomass/methane is converted to synthesis gas through the gasification process with limited oxygen/air and then converted to methanol followed by dehydration to DME. DME is produced through a chemical process, as shown in Equations 2.28 through 2.31 (Troy et al., 2006).

$$CO + 2H_2 \rightarrow CH_3OH; \Delta H^0 = -90.3 \frac{kJ}{mol} \tag{2.28}$$

$$2CH_3OH \rightarrow CH_3OCH_3 + H_2O; \Delta H^0 = 23.4 \frac{kJ}{mol} \tag{2.29}$$

$$H_2O + CO \rightarrow H_2 + CO_2; \Delta H^0 = 40.9 \frac{kJ}{mol} \tag{2.30}$$

$$3H_2 + 3CO \rightarrow CH_3OCH_3 + CO_2; \Delta H^0 = 258.6 \frac{kJ}{mol} \tag{2.31}$$

The economic growth rate of the gross domestic products of China and India are the highest in the world. The IEA envisages that India has the potential to produce second-generation biofuels (biosynthetic natural gas, biomass to liquid (BTL) biodiesel and bioethanol), which can be produced from feedstocks, including rice straw, wheat straw, corn stalk and stems, sorghum stalk, cotton stalk, bagasse, rice husk, and peanut shell, whereas China could produce biofuels (biosynthetic natural gas, BTL biodiesel and bioethanol) from rice straw, corn (stalk, stem), wheat straw, cotton stalk, oil crops, logging residue, rice husk, corncob, bagasse, and saw mill residue. But other biofuels, including DME, methanol, and hydrogen get also more attention. Therefore, biofuels mentioned above such as methanol, ethanol, butanol, biogas, enriched biogas, hydrogen, biodiesel, F-T diesel, and DME are discussed in upcoming chapters.

Summary

Biofuels (methanol, ethanol, butanol, biogas, enriched biogas, hydrogen, biodiesel, F-T diesel, and DME) can be produced from biomass. These biofuels have better ignition quality (cetane number) and antiknocking indexes (octane number) than those of petroleum fuels (gasoline and diesel). Because biofuels such as methanol, ethanol, butanol, enriched biogas, and hydrogen have higher octane numbers, they can be used in spark-ignition engines. Engines with higher octane number fuels have a better opportunity to improve thermal efficiency by optimizing their compression ratios. Biodiesel, F-T diesel, and DME have higher cetane numbers (better ignition quality) and they are suitable for compression-ignition engines. The transient performance of a compression-ignition engine with biofuels would be improved, along with other benefits, such as better cold-start ability and transient emissions reduction.

References

Ainara, A., Pérez-Uriarte, P., Sánchez-Contador, M., Ereña, J. et al. (2016), Direct synthesis of dimethyl ether from syngas on CuOZnOMnO/SAPO-18 bifunctional catalyst, *International Journal of Hydrogen Energy*, 41(40), 18015–18026.

Anselm Eisentraut (2010), Sustainable production of second generation biofuels—Potential and perspectives in major economies and developing countries, International Energy Agency (IEA).

Arcoumanis, C. (2000), The Second European Auto-Oil Programme (AOLII). European Commission, Vol. 2. Alternative Fuels for Transportation.

Arcoumanis, C., Bae, C., Crookes, R. and Kinoshita, E. (2008), The potential of di-methyl ether (DME) as an alternative fuel for compression-ignition engines: A review, *Fuel*, 87, 1014–1030.

Argun, H., Kargi, F. and Kapdan, I. K. (2009), Microbial culture selection for bio-hydrogen production from waste ground wheat by dark fermentation, *International Journal of Hydrogen Energy*, 34, 2195–2200.

Balat, H. and Kirtay, E. (2010), Hydrogen from biomass present scenario and future prospects, *International Journal of Hydrogen Energy*, 35, 7416–7426.

Biradar, C. H., Subramanian, K. A. and Dastidar, M. G. (2014), Production and fuel quality upgradation of pyrolytic bio-oil from Jatropha Curcas de-oiled seed cake, *Fuel*, 119, 81–89.

Chmielniak, T. and Sciazko, M. (2003), Co-gasification of biomass and coal for methanol synthesis, *Applied Energy*, 74, 393–403.

Das, D. and Veziroğlu, T. N. (2001), Hydrogen production by biological processes: A survey of literature, *International Journal of Hydrogen Energy*, 26(1), 13–28.

Das, D. and Veziroglu, T. N. (2008), Advances in biological hydrogen production process, *International Journal of Hydrogen Energy*, 33, 6046–6057.

Das, D., Khanna, N. and Veziroglu, T. N. (2008), Recent developments in biological hydrogen production processes, *Chemical Industry and Chemical Engineering Quarterly*, 14, 57–67.

Demarrias, A. and Demirbas, M. F. (2011), Importance of algae oil as a source of biodiesel, *Energy Conversion and Management*, 52, 163–170.

Ditzig, J., Liu, H. and Logan, B. E. (2007), Production of hydrogen from domestic wastewater using a bio electrochemically assisted microbial reactor (BEAMR), *International Journal of Hydrogen Energy*, 32, 2296–2304.

Dong, Y. and Steinberg, M. (1997), Hynol—An economical process for methanol production from biomass and natural gas with reduced CO2 emission, *Hydrogen Energy*, 22, 971–977.

ETBE, European biofuels, http://www.biofuelstp.eu/etbe.html (accessed on 1/15/2017).

Fabiano, B. and Perego, P. (2002), Thermodynamic study and optimization of hydrogen production by *Enterobacter aerogenes*, *International Journal of Hydrogen Energy*, 27, 149–156.

Fang, H. H. P. and Liu, H. (2002), Effect of pH on hydrogen production from glucose by a mixed culture, *Bioresource Technology*, 82, 87–93.

Galindo, C. P. and Badr, O. (2007), Renewable hydrogen utilisation for the production of methanol, *Energy Conversion and Management*, 48, 519–527.

Gill, S. S., Tsolakis, A., Dearn, K. D. and Rodríguez, F. J. (2011), Combustion characteristics and emissions of Fischer–Tropsch diesel fuels in IC engines, *Progress in Energy and Combustion Science*, 37(4), 503–523.

Güllü, D. and Demirbas, A. (2001), Biomass to methanol via pyrolysis process, *Energy Conversion and Management*, 42, 1349–1356.

Hasegawa, F., Yokoyama, S. and Imou, K. (2010), Methanol or ethanol produced from woody biomass: Which is more advantageous? *Bioresource Technology*, 101, 109–111.

Hawkes, F. R., Dinsdale, R., Hawkes, D. L. and Hussy, I. (2002), Sustainable fermentative hydrogen production: Challenges for process optimization, *International Journal of Hydrogen Energy*, 27, 1339–1347.

Hawkes, F. R., Hussy, I., Kyazze, G., Dinsdale, R. et al. (2007), Continuous dark fermentative hydrogen production by mesolphilic microflora: Principles and progress, *International Journal of Hydrogen Energy*, 32, 172–184.

Jingura, R. M. and Rutendo, M. (2009), Optimization of biogas production by anaerobic digestion for sustainable energy development in Zimbabwe, *Renewable and Sustainable Energy Reviews*, 13(5), 1116–1120.

Keshav, T. R. and Basu, S. (2007), Gas-to-liquid technologies: India's perspective, *Fuel Processing Technology*, 88(5), 493–500.

Kumabe, K., Fujimoto, S., Yanagida, T., Ogata, M. et al. (2008), Environmental and economic analysis of methanol production process via biomass gasification, *Fuel*, 87, 1422–1427.

Kumar, N. and Das, D. (2000), Enhancement of hydrogen production by *Enterobacter cloacae* IIT-BT 08, *Process Biochemistry*, 35, 589–593.

Lercher, J. A., Bitter, J. H., Steghuis, A. G., Ommen, J. G. Van, Seshan, K. (1999), *Environmental Catalysis*, Imperial College Press, London, 103–126.

Liu, H., Grot, S., Logan, B. E. (2005), Electrochemically assisted microbial production of hydrogen from acetate, *Environmental Science & Technology*, 39, 4317–4320.

Melis, A. and Melnicki, M. R. (2006), Integrated biological hydrogen production, *International Journal of Hydrogen Energy*, 31, 1563–1573.

Melnicki, M. R., Bianchi, L., Philippis R. De and Melis, A. (2008), Hydrogen production during stationary phase in purple photosynthetic bacteria, *International Journal of Hydrogen Energy*, 33, 6525–6534.

MTBE, American cancer society, http://www.cancer.org/cancer/cancercauses/othercarcinogens/pollution/mtbe (accessed on 1/15/2017).

Nakagawa, H., Harada, T., Ichinose, T., Takeno, K. et al. (2007), Bio-methanol production and CO2 emission reduction from for age grasses, trees, and crop residues, *Japan Agricultural Research Quarterly: JARQ Agriculture Resources*, 41, 173–80.

Nautiyal, P., Subramanian, K. A. and Dastidar, M. G. (2014), Production and characterisation of biodiesel from algae, *Fuel Processing Technology*, 120, 79–88.

Ni, Y. and Sun, Z. (2009), Recent progress on industrial fermentative production of acetone–butanol–
    ethanol by *Clostridium acetobutylicum* in China, *Applied Microbiology and Biotechnology*, 83,
    415–423.
Ogbonna, J. G., Soejima, T. and Tahala, H. (1999), An integrated solar and artificial light system for
    internal illumination of photo bioreactors, *Journal of Biotechnology*, 70, 289–297.
Rajesh Kumar, B. and Saravanan, S. (2016), Use of higher alcohol biofuels in diesel engines: A review,
    Renewable & Sustainable Energy Reviews, 60, 84–115.
Rasskazchikova, T. V., Kapustin, V. M. and Karpov, S. A. (2004), Ethanol as high–octane additive to
    automotive gasolines. Production and use in Russia and abroad, *Chemistry and Technology of
    Fuels and Oils*, 40(4), 203–210.
Renó, M. L. G., Lora, E. E. S., Palacio, J. C. E., Venturini, O. J. et al. (2011), LCA (life cycle assessment)
    of the methanol production from sugarcane bagasse, *Energy*, 36, 3716–3726.
Sahibzada, M., Chadwick, D. and Metcalfe, I. S. (1997), Methanol synthesis from CO2/H2 over
    Pd-promoted Cu/ZnO/Al2O3 catalysts: Kinetics and deactivation, *Studies in Surface Science
    and Catalyst*, 107, 29–34.
Sarathy, S. M., Oßwald, P., Hansen, N. and Kohse-Höinghaus, K. (2014), Alcohol combustion chem-
    istry, *Progress in Energy and Combustion Science*, 44, 40–102.
Semelsberger, T. A., Borup, R. L. and Greene, H. L. (2006), Dimethyl ether (DME) as an alternative
    fuel, *Journal of Power Sources*, 156(2), 497–511.
Shamsul, N. S., Kamarudin, S. K., Rahman, N. A. and Kofli, N. T. (2014), An overview on the produc-
    tion of bio-methanol as potential renewable energy, *Renewable and Sustainable Energy Reviews*,
    33, 578–588.
Sharma, Y. C. and Singh, V. (2017), Microalgal biodiesel: A possible solution for India's energy secu-
    rity, *Renewable and Sustainable Energy Reviews*, 67, 72–88.
da Silva Trindade, W. R. and dos Santos, R. G. (2017), Review on the characteristics of butanol, its
    production and use as fuel in internal combustion engines, *Renewable and Sustainable Energy
    Reviews*, 69, 642–651.
Subramanian, K. A., Mathad Vinaya, C., Vijay, V. K. and Subbarao, P. M. V. (2013), Comparative
    evaluation of emission and fuel economy of an automotive spark ignition vehicle fuelled
    with methane enriched biogas and CNG using chassis dynamometer, *Applied Energy*, 105,
    17–29.
Taylor, C. E., and Noceti, R. P. (2000), New developments in the photo catalytic conversion of meth-
    ane to methanol, *Catalysis Today*, 55, 259–267.
Themelis, N. J. and Ulloa, P. A. (2007), Methane generation in landfills, *Renewable Energy*, 32(7),
    1243–1257.
Wakayama, T. and Miyake, J. (2002), Light shade bands for the improvement of solar hydrogen pro-
    duction efficiency by *Rhodobacter sphaeroides* RV, *International Journal of Hydrogen Energy*, 27,
    1495–1500.
Weimer, T., Schaber, K., Specht, M. and Bandi, A. (1996), Methanol from atmospheric carbon dioxide:
    A liquid zero emission fuel for the future, *Energy Conversion and Management*, 37, 1351–1356.
Xu, S. W., Lu, Y., Li, J., Jiang, Z. Y. and Wu, H. (2006), Efficient conversion of CO2 to methanol cata-
    lysed by three de-hydrogenises co-encapsulated in an alginate-silica (ALG-SiO2) hybrid gel,
    Industrial Engineering and Chemical Research, 45, 4567–4573.
Zhen, X. and Wang, Y. (2015), An overview of methanol as an internal combustion engine fuel,
    Renewable and Sustainable Energy Reviews, 52, 477–493.

3

Biofuel Quality for Internal Combustion Engines

3.1 Properties of Gasoline Fuel and Their Definitions

3.1.1 Octane Number

The octane number indicates the antiknocking property of fuels for spark-ignition engines. The octane number is a dimensionless number and is a measure of gasoline's ability to resist autoignition. Autoignition is an undesirable preignition in spark-ignition engines. It is directly related to the autoignition temperature of fuels. A high octane number indicates a qualitatively high autoignition temperature whereas a high cetane number indicates a low autoignition temperature and vice versa, but the value of the octane number does not equal the value of the autoignition temperature. A lower octane number of fuel may cause autoignition before spark discharge, which is called preignition phenomenon, whereas fuel causing autoignition due to compression of the end charge by flame speed is called postignition phenomenon.

The charge (air-fuel mixture) causes autoignition in spark-ignition engines if the octane number/ignition temperature of fuel is less than the required level. Autoignition of a charge results in a drastic increase in a higher rate of pressure rise and abrupt change in fluctuation of in-cylinder pressure. The in-cylinder temperature and pressure in a combustion chamber with autoignition are relatively too high, resulting in severe engine knock, more vibration, and audible sound. The high intensity of knock would damage engine components such as pistons and valves. If the octane number is high, the knocking problem is minimized or can be eliminated. The compression ratio of spark-ignition engines can be increased with an increase in octane number and engine operation at an increased compression ratio would yield higher thermal efficiency or fuel economy.

The octane number is defined as the antiknocking index that is calculated using the data obtained from the cooperative fuel research (CFR) engine. The CFR engine is operated per the American Society for Testing and Materials (ASTM) procedure for finding out the octane number of any fuel. The octane number is classified into two different numbers: RON and MON. RON indicates the performance of a spark-ignition engine with part load and less intensity of knock, whereas MON relates to an engine with a severe knocking condition. Fuel sensitivity is the difference between RON and MON:

$$\text{Fuel sensitivity} = \text{RON} - \text{MON}$$

Biofuels such as methanol, ethanol, butanol, biogas, and hydrogen have a higher octane number (>85) than that of a base gasoline engine (85–95). Therefore, a biofueled spark-ignition engine can operate with a higher compression ratio and thus with higher thermal efficiency with less emissions. The reactant's pressure and temperature increase with increasing compression ratio, resulting in an increase in the chemical reaction rate of fuel with air. The heat release rate is higher with a higher compression ratio or the quality of heat increases that results in higher thermal efficiency. Since the in-cylinder temperature is higher, emissions such as CO, HC, and particulate matter, except for NOx, would be less.

3.1.2 Sulfur

Sulfur is naturally embedded with crude hydrocarbon. Sulphur embedded with minerals and oil stored underground does not have any negative environmental impact because it does not interact with the atmosphere, but if sulfur-embedded hydrocarbon is combusted by heat engines or internal combustion engines, the sulfur is oxidized into sulfuric acid, sulfate, or SOx emission. The SOx emission produced by the combustion system would corrode the engine's components and the emission interface's environment negatively. The SOx component is dissolved in the water vapor in clouds in the atmosphere and it comes back to earth as acid rain, which is harmful to aquatic species because acid rain decreases the pH level of water body systems (lake, pond, river, and ocean).

If the sulfur is not removed during the refining process, it will emerge as SOx emission from engines/vehicles. Sulfur-based emission inhibits functioning of the catalyst in catalytic converters, resulting in poor conversion efficiency, catalyst poisoning, and harmful sulfur-based emissions (due to better oxidation (negative advantage) of sulfur species by catalysts). Sulfur can affect sensors, including oxygen sensors, which are fitted in the engine's exhaust gas system. Reduction in sulfur is mandatory in current and future fuel quality worldwide and very important for better functioning of after treatment device fitted in vehicles as well as maintaining a sustainable environment.

3.1.3 Olefins

The molecular structure of olefins is

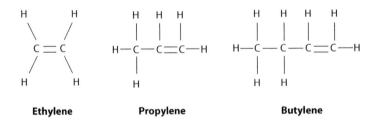

Ethylene   **Propylene**   **Butylene**

Hydrocarbon with carbon – carbon double bond (unsaturated HC)

Chemical formula – (C_nH_{2n}), n ≥ 2

Olefin-structured compound forms mainly during the refinery process while upgrading crude oil to the desired fuel. Hydrocarbon with an olefin structure generally has a higher octane number and has a positive impact on reduction of hydrocarbon emissions. However, fuel with this structure has numerous disadvantages. For example, olefins in fuel may lead to deposits on intake valves and other engine components. Current fuel quality norms indicate that olefin-structured hydrocarbon should be decreased. Subsequently, future fuel quality norms may dictate that the olefin component be less or nil in fuel.

3.1.4 Aromatics

The molecular structure of aromatic hydrocarbons are

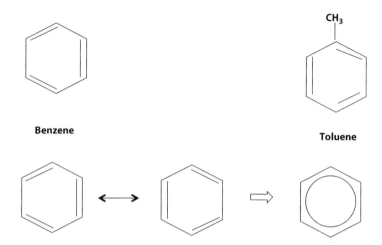

Benzene

Toluene

A fuel molecule with aromatic structure contains at least one benzene ring. In general, aromatics are ring structures because the bond between hydrogen and carbon are a branched ring structure. Fuel with a ring structure is generally difficult to break down, resulting in a high octane number and this structured fuel needs an external ignition aid to break the bond for initiation of ignition of the air-fuel charge. For example, isobutane has a higher octane number than that of N-butane as the bond strength of the isomerization (iso-) structure is higher than that of normal structure (N-). Similarly, the octane number of iso-octane is higher than that of N-octane. Aromatic structured molecules have high-energy density. However, the aromatic content in fuel can increase engine deposits and increase tailpipe emissions, including polycyclic aromatic hydrocarbon (PAH) and benzene emissions. The level of aromatic-structured molecules in fuel must be lower in order to meet present and future quality norms.

3.1.5 Benzene

The molecular structure of benzene is

Benzene has a high octane number and can act as an octane booster for gasoline fuels. However, it is not advisable to blend it in gasoline due to harmful emissions from the engine. The benzene level in fuel must be reduced or eliminated in order to meet fuel quality and vehicular emissions norms.

3.1.6 Vapor Pressure

The vapor pressure of gasoline should be controlled seasonally because engines need volatility corresponding to different ambient temperatures. The vapor pressure must be at a level that will avoid the possibility of hot-fuel handling problems such as vapor lock. Vapor pressure is also important at lower temperatures for ease of starting and part load performance. Therefore, the vapor pressure in a fuel must be within a range that allows for satisfactory performance of engines.

3.1.7 Distillation

Distillation properties of fuel/gasoline indicates the temperature at which fuel vapor forms. Distillation properties can also be defined as either a set of T points (T50 is the temperature at which 50% of the gasoline distills) or E points (E100 is the percentage of a gasoline distilled at 100°C). T10 indicates the ease of starting an engine; if the vapor is less than the desired level at a corresponding temperature, the engine would experience cold-start problems. If T10 is higher than the required level, there will be too much vapor, which may lead to starting problems. Distillation can be used for predicting the engine's cold-start and warm-up performance. The distillation point T50 is important for engine function at part load. T90 indicates that 90% of fuel is being converted to vapor corresponding to the available temperature in the cylinder. Note that the average temperature of an engine is above 1500°C during combustion and the exhaust temperature is in the range of approximately 300°C to 500°C. The localized temperature distribution is different in the combustion chamber. The temperature of charge/burned gas in the nearby cylinder wall is lower due to heat transfer loss to the coolant through the cylinder wall. Therefore, the fuel has to be a vapor at the engine's temperature; otherwise, fuel droplets may be partially oxidized or deposited on the combustion chamber walls or piston surfaces. T90 indicates the possibility of deposits on the engine components, and therefore, the distillation characteristics are important for qualitatively predicting a cold start, cruising speed performance, and deposits.

3.1.8 Lead

Tetra-ethyl lead is an inexpensive octane booster or enhancer for gasoline fuel. However, it will negatively affect the function of a catalytic converter or any catalyst aftertreatment device in an engine. Lead content in fuel reduces the function and life of sensors, including the oxygen sensor used in exhaust systems. Unleaded gasoline is mandatory for catalyst-equipped vehicles. A lead-free fuel is also essential for human health. Due to the above reasons, most countries use unleaded gasoline in their vehicle fleets.

The important fuel quality parameters of gasoline fuel include RON, MON, density, distillation characteristics, Reid vapor pressure (RVP), vapor lock index, copper strip corrosion, water tolerance of gasoline-alcohol blends, oxidation stability, and benzene, olefin, aromatic, sulfur, and lead contents. A spark-ignition engine fuel must meet required fuel quality norms. For example, the density of fuel must be in the prescribed range; otherwise, it may cause leaking of fuel in the injection system or poor mixing with air because heavier molecules are mostly high density. The octane number of fuel must be more than the prescribed level. If the octane number is less than the allowed value, the engine will operate with knocks, resulting in decreasing engine life, stalling the engine operation, or deterioration of the engine's performance and emissions. If it is higher than the prescribed value, the engine will operate smoothly without knocking. The increasing octane number is a complex refinery process and incurs additional cost to the fuel. The important point to note is that engine technology has to be upgraded—for example, increasing compression ratio—for the benefits of a higher octane number to be realized. Therefore, engines with the same capacity will yield different fuel economy based on the value of the octane number. The octane number is decided based on consultations with stakeholders, including oil companies and the original engine manufacturers (OEMs). If a fuel does not meet the agreed-on fuel quality, it cannot be used in engines; otherwise, the engine's performance and emissions will deteriorate and may stall, fail, or its life will be decreased. For example, the specification of commercial gasoline fuel being used in India is given in Table 3.1. In India, the gasoline fuel shall meet the specification listed in the table. For example, RON and MON of the fuel shall be the minimum of 91 and 81, respectively; otherwise, the fuel does not meet the required quality and is not acceptable as a vehicular fuel. It may be noted that fuel quality norm may vary from country to country based on the kind of particular vehicle technology and emission norms.

TABLE 3.1

Specification of Commercial Gasoline Fuel in India

Serial Number	Properties		Gasoline Requirements
1	Density at 15°C, kg/m³	–	720–775
2	Distillation		
	• Recovery up to 70°C (% vol)		10–55 (summer)
			10–58 (other months)
	• Recovery up to 100°C (% vol)	Min	40–70
	• Recovery up to 150°C (% vol)	Min	75
	• Final boiling point, °C	Max	210
	• Residue (% vol)		2
3	RON	Min	91
4	MON	Min	81
5	Olefin content, % volume	Max	21
6	Aromatic content, % volume	Max	35
7	Lead content, g/l	Max	0.005
8	Sulfur content, total, mg/kg	Max	10
9	Reid vapor pressure at 38°C, kPa	Max	67
10	Oxygenate content		
	• Methanol, % volume	Max	3
	• Ethanol, % volume	Max	10
	• Isopropyl alcohol, % volume	Max	10
	• Isobutyl alcohol, % volume	Max	10
	• Tertiary-butyl alcohol, % volume	Max	7
	• Ethers containing five or more carbon atoms per molecule, % volume	Max	15
	• Other oxygenates, % volume	Max	8

Source: The Gazette of India, Part II, Section 3, Subsection (i), Ministry of Road Transport and Highways Notification, September 16, 2016, New Delhi.

3.2 Properties of Diesel Fuel and Their Definition

3.2.1 Cetane Number

The cetane number is a dimensionless ignition parameter that indicates the ignition quality of fuel. The ignitability of fuel increases with an increase in the cetane number. The ignition delay period, which is defined as the difference between the start of combustion and the start of dynamic injection timing, decreases with an increase in the cetane number of fuel. The ignition delay period is an important parameter as it influences rate of pressure rise. If the cetane number is less than the desired level, the ignition delay period increases, leading to more accumulation of injected fuel during the ignition delay period, which in turns results in combustion with knock. The cetane number indicates the qualitative performance of the engine in terms of whether it will be able to (1) start without problems during the winter or in cold climate conditions, (2) reduce exhaust emissions (transient as well as steady-state condition), and (3) reduce combustion noise. A higher cetane number generally enables control of the ignition delay and combustion stability and is preferable for modern or advanced engines that work with technologies such as exhaust gas recirculation (EGR), homogeneous charge compression ignition (HCCI), and reactivity controlled compression ignition (RCCI). Thus, if the fuel has a higher cetane number, the engine can effectively be calibrated for incorporation of advanced emission reduction technologies with less noise and better engine performance.

The cetane number is generally measured using a CFR engine or Ignition Quality Tester (IQT) as per ASTM D613 test procedure. The cetane index (ASTM D4737) is also calculated using measured physicochemical properties of fuel (fuel density and distillation temperatures). The cetane number is directly measured using the research engine/IQT system whereas the cetane index is a correlated parameter obtained from fuel properties. The cetane number of alternative liquid fuels (with more of a paraffin structure) can be measured using the engine but calculation of the cetane index is limited to certain fuels as it may not be calculated by correlating it with properties of alternative fuels or biofuels. For example, the diesel index is calculated using aniline point and specific gravity. The aniline point correlates with paraffin or aromatics and it is higher with paraffin compounds and lower with aromatics and vice versa. However, diesel index type correlation cannot be derived for biodiesel or F-T diesel because these fuels do have aromatic compounds.

3.2.2 Density and Viscosity

Density and viscosity are the main functions of fuel spray characteristics such as breakup length, penetration distance, Sauter mean diameter, and air entrainment. These properties directly affect the atomization of injected fuel. Density influences the mass flow rate of fuel, whereas the viscosity of fuel affects atomizing as well as flow ability by injection system. If fuel viscosity is too high, the input power requirement of the fuel injection system would increase. If viscosity is lower than the desired level, fuel leakage in the pump may be high. Therefore, the density and viscosity of fuels must be in the desired level range.

3.2.3 Sulfur

Particulate matter decreases with a decrease in the sulfur content in fuel. Sulfur levels should be at ultra for vehicles fitted with aftertreatment devices; otherwise, the exhaust aftertreatment system would be ineffective due to sulfur poisoning.

3.2.4 Aromatics

Molecules with an aromatic structure contain at least one benzene ring. The aromatic-structured molecule during combustion influences the formation of particulate and PAH emissions. The temperature of an in-cylinder may increase due to thermal pyrolysis of fuel during combustion. The aromatic component may release more energy at higher temperature that may result in higher NOx emission as compared to other components like paraffin.

3.2.5 Distillation Characteristics

Distillation properties indicate a vaporizing tendency at given temperatures. The distillation curve can be divided into three regions:

- The light end (beginning point), which affects start ability.
- Middle distillation point, which refers to vaporization of 50% fuel (T50) corresponding to an engine's temperature, influences the cruising performance of vehicles.
- The end distillation points (T90, T95) refer to vaporization of 90% or 95% corresponding to an engine's temperature. These properties are used to assess qualitatively the degree of mixing process of air and fuel, coke formation into holes of fuel injectors, and soot emission.

Engine performance can be qualitatively predicted using the distillation characteristics. The distillation characteristics at the heavy end are well correlated with engine performance however a lower boiling range is inconclusive. The distillation properties of a fuel indicate the vaporizing tendency corresponding to an engine's temperature. The required distillation properties will ensure that the fuel state will completely be changed from liquid to vapor state before or during combustion in an engine.

3.2.6 Cold-Flow Characteristics

Cold-flow characteristics deal with the ability of fuel to flow in cold climate conditions. These characteristics are generally described using cloud point, cold filter plugging point, and low-temperature flow test (LTFT).

- *Cloud Point*: The fuel looks hazy at cold temperatures and is relatively less transparent. Heavier molecules may start forming wax during the winter season.
- *Cold Filter Plugging Point* (CFPP): The fuel is passed to the filter, which filters foreign particles that are larger than its porous hole size. If the fuel consistency is like wax or is hazy, the fuel will not effectively flow into the filter element. The threshold lowest temperature at which a filter can function well is the CFPP. A low-temperature flow test (LTFT) for fuel is generally done at more severe temperatures for assessing cold-flow characteristics.

3.2.7 Lubricity

Lubricity of fuel is an important property for a fuel-injection system. For example, an injection system does not need any external lubrication because diesel fuel itself acts as a self-lubricant. The lubricity of diesel is less with sulfur removal. This problem can be overcome by multifunctional additives; however, these measures may lead to additional problems of deposit on engine components and higher cost of fuel. In the case of biodiesel, it has a higher lubricity property whereas DME and F-T diesel have less lubricity. A high-frequency reciprocating rig (HFRR) test is used for quantifying a fuel's lubricity.

3.2.8 Particulate Contamination

The nozzle-hole diameter of a common-rail direct injection (CRDI) diesel engine is much smaller compared to earlier conventional engines (non-CRDI). If any foreign particle is present in the fuel, it severely affects the operation of the fuel-injection system, the functioning of physical filter, and so forth.

3.2.9 Flash Point

The flash point is one of most important fuel properties for safe handling of fuel during storage, transportation, and dispensing. The flash point is defined as the temperature at which flammable vapor can form. For example, if the flash point of diesel is about 35°C, the liquid fuel can be in vapor form if the surrounding temperature is above 35°C. Flammable vapor can also catch fire if any external ignition source is present.

The important physico-chemical properties of diesel fuel include cetane number, density, viscosity, distillation characteristics, sulfur, flash point, PAHs, total contaminants, carbon residue, water content, lubricity (wear scar diameter (WSD) at 60°C), ash content, CFPP, oxidation stability, and copper strip corrosion. For example, the specification of commercial diesel fuel in India is given in Table 3.2. It can clearly be observed from the table that the value of the limit of each fuel property is specified. For example, the cetane number of diesel must be more than 51 or the fuel must not be sold or used in vehicles and is a violation of fuel quality norms. The cetane number value may differ from country to country based on their oil and vehicles' fleet context. The density of fuel mentioned in

TABLE 3.2

Specification of Commercial Diesel Fuel in India

Serial Number	Properties		Requirements
1	Density at 15°C, kg/m³	Max	845
2	Distillation (T95), °C	Max	360
3	Total Sulphur, mg/kg	Max	10
4	Cetane No	Min	51
5	Cetane Index	Min	46
6	Flash Point, °C	Min	35
7	Kinematic Viscosity at 40°C, cSt		2.0–4.5
8	Water content, mg/kg	Max	200
9	Ash, % mass	Max	0.01
10	Total contaminations, mg/kg	Max	24

Source: The Gazette of India, Part II, Section 3, Subsection (i), Ministry of Road Transport and Highways Notification, September 16, 2016, New Delhi.

the table must be a maximum of 845 kg/m³. In this case, if a fuel has less than 845 kg/m³, a fuel-injection system may have the problem of more leakage. In contrast, if the density is higher than the specified limit, fuel spray characteristics will be poor, especially larger breakup length, less spray cone angle, and larger Sauter mean diameter (SMD), because density is one of the main parameters for spray characteristics of compression-ignition engines. Similarly, more leakage will occur in an injection system if viscosity is lower; otherwise, spray characteristics will be poor with high viscosity. If a fuel does not meet the desired fuel quality norms, the fuel must not be used in engines; otherwise, performance and emissions characteristics will deteriorate. In this case, if poor-quality fuel is used in a vehicle, this may be considered as violating the norms and federal rules and procedures being followed by a particular country.

The most important biofuels for spark-ignition engines and compression-ignition engines are listed next. The selection of the biofuels are mainly based on their feedstock resource potential and practical feasibility to implement in vehicles.

3.3 Biofuels for Spark-Ignition Engines

A selection of a fuel for spark-ignition engines is mainly based on fuel's octane number for ensuring combustion without knock. The following biofuels are more suitable for SI engines as they have a higher octane number than gasoline.

- Methanol
- Ethanol
- Butanol
- Biogas
- Hydrogen

3.4 Biofuels for Compression-Ignition Engines

A selection of a fuel for compression-ignition engines is mainly based on fuel's cetane number (ignition quality) for ensuring combustion without knock as well as smooth engine running. The following biofuels are more suitable for CI engines as they have a higher cetane number than diesel.

- Biodiesel
- F-T diesel
- DME

The fuel quality of the above biofuels is briefly discussed next.

3.5 Methanol

The molecular structure of methanol is

$$
\begin{array}{c}
\text{H} \\
| \\
\text{H} - \text{C} - \text{OH} \\
| \\
\text{H}
\end{array}
$$

Methanol

Methanol is an oxygenated fuel and has the lowest carbon number in the alcohol fuel family. The physico-chemical properties of methanol are given in Table 3.3.

Methanol has a higher octane number and hence it is a better fuel for spark-ignition engines. The octane number of methanol is higher than gasoline, so engines can be operated at a higher compression ratio to improve their thermal efficiency. The oxygen content in this fuel would lead to better oxidation of CO, HC, and particulate matter in spark-ignition engines compared to base gasoline. As the calorific value of methanol is less than almost half of the calorific value of gasoline, the power and torque of an engine with methanol may be less than with base gasoline. In order to maintain the same power output, a fuel-injection system needs to be tuned to supply almost double the fuel quantity to achieve the same power compared to gasoline fuel; otherwise, the power output will drop. The change in NOx emission depends on the equivalence ratio and operating conditions of the engine. Methanol has a higher latent heat of vaporization resulting in charge cooling and hence higher volumetric efficiency. On the other hand, higher latent heat of vaporization may lead to poor atomization of fuels at low atmospheric temperature.

TABLE 3.3

Physico-Chemical Properties of Methanol

Fuel Property	Units	Methanol
Formula	–	CH_3OH
Molecular weight	g.mol	32
Oxygen content	–	50%
Low calorific value	MJ/kg	19.66
High calorific value	MJ/kg	22.3
Stoichiometric air/fuel ratio	–	6.45
Boiling point	°C	64.8
Flash point	°C	11
Autoignition temperature	°C	465
Motor octane number	–	88.6
Research octane number	–	108.7
Latent heat of vaporization	kJ/kg	1109
Viscosity (at 20°C)	cP	0.6

Source: Zhen, X. and Wang, Y. (2015), An overview of methanol as an internal combustion engine fuel, *Renewable and Sustainable Energy Reviews*, 52, 477–493.

3.6 Ethanol

The molecular structure of ethanol is

```
        H   H
        |   |
H ——— C —— C ——— OH
        |   |
        H   H
```

Ethanol

Ethanol is considered an octane booster because the octane number of gasoline can further be increased if ethanol at any percentage is blended with gasoline. The physico-chemical properties of ethanol are given in Table 3.4.

Ethanol has similar physico-chemical properties to those of methanol. The high octane number of methanol fuel could reduce the probability of engine knocking. An engine's thermal efficiency could be enhanced by increasing the engine's compression ratio because ethanol has a higher octane number. When ethanol is blended with gasoline at different percentages, the octane number increases but the calorific value of the fuel decreases. This necessitates supplying a greater quantity of fuel per cycle compared to neat gasoline. It is important to note that both methanol and ethanol fuels do not contain sulfur, olefins, or aromatics. CO and HC emissions decrease with ethanol-blended gasoline compared to neat gasoline. This is mainly due to fuel-embedded oxygen that ensures complete combustion of fuel. The NOx emission trend is dependent on the engine operating condition and equivalence ratio. The higher latent heat of vaporization may have both positive and negative effects on performance and emissions. At high loads, the volumetric efficiency would increase due to the charge cooling effect of an ethanol-gasoline mixture. This would help to increase the power and torque output of the engine.

TABLE 3.4

Physico-Chemical Properties of Ethanol

Property	Units	Ethanol	E5	E10	E25	E50
Ethanol content	% vol, min	99.8 (Sweden), 99.6 (Poland), 99.3 (Ukraine), 92.1 (United States), 99.3 (Brazil)	5.01	10.82	26.42	49.76
Density	kg/m³, max	790 (Sweden), 791.5 (Brazil)	751.9	756	761.3	772.4
Calorific value	MJ/kg	28	41.99	41.24	40.83	33.72
Motor octane number	–	92	86.8	84.8	86.4	88.1
Research octane number	–	108	95.0	96.6	100.5	103.2

Source: Bielaczyc et al. (2013), An examination of the effect of ethanol–gasoline blends' physicochemical properties on emissions from a light-duty spark ignition engine, *Fuel Processing Technology*, 107, 50–63; Rutz et al. (2006), Project: Biofuel Marketplace (EIE/05/022/SI2.420009), *Overview and Recommendations on Biofuel Standards for Transport in the EU*, WIP Renewable Energies, Germany.

3.7 Butanol

The molecular structure of butanol is

Butanol

The physico-chemical properties of butanol are given in Table 3.5.

Butanol has the highest calorific value compared to the other two alcohol fuels. This is advantageous because power drop may be less with butanol compared to other alcohol fuels. The flash point of butanol is higher compared to ethanol and methanol fuels. This allows butanol to be stored and transported in a less complex way. The major hurdle in the use of butanol is poor vaporization characteristics resulting in improper air-fuel mixture formation. Butanol is less volatile compared to ethanol or gasoline. The system needs higher injection pressure or cold-start devices such as glow plugs to improve the mixture formation. Butanol is more compatible with the materials used in vehicle fuel systems compared to ethanol. Like other alcohol fuels, fuel-embedded oxygen in butanol helps to improve combustion and reduce CO and HC emissions in spark-ignition engines. The performance of the spark-ignition engine with higher

TABLE 3.5

Physico-Chemical Properties of Butanol

Fuel Property	Units	n-Butanol
Molecular formula	–	C_4H_9OH
Molecular weight	g/mol	74.11
Cetane number	–	25
Motor octane number	–	85
Research octane number	–	98
Oxygen content	% weight	21.6
Density (at 20 °C)	kg/m³	808
Autoignition temperature	°C	385
Flash point	°C	35
Lower heating value	MJ/kg	33.1
Boiling point	°C	117.7
Stoichiometric ratio	–	11.21
Latent heating (at 25°C)	kJ/kg	582
Viscosity (at 40°C)	mm²/s	2.63

Source: da Silva Trindade, W. R. and dos Santos, R. G. (2017), Review on the characteristics of butanol, its production and use as fuel in internal combustion engines, *Renewable and Sustainable Energy Reviews*, 69, 642–651. (Reprinted with permission.)

butanol blends is almost similar to that of other alcohol blends. The volumetric consumption of alcohol fuels is higher with butanol due to lower energy content compared to gasoline fuel, but is better with butanol fuel compared to methanol and ethanol fuels.

3.8 Biogas and Enriched Biogas

Biogas can be produced from waste biomass including seed cake, cattle dung, human waste (night soil), algae residue, municipal waste, agricultural waste, and wastewater from a treatment plant. Biogas plant has become almost mandatory in many industries, including wastewater treatment plants, because biogas production from effluent/waste in industries will lead to a reduction of the biological oxygen demand (BOD) level in waste; otherwise, higher BODs at unacceptable levels in effluents may lead to waste disposal problems such as dumping into aquatic systems. In addition, biogas generated from waste resources can be utilized in combustion engines for power generation; otherwise, the emissions (mainly methane) from these resources will release into the atmosphere. As methane has a higher global warming potential (GWP) than carbon dioxide, emission release into the atmosphere would negatively contribute to global warming and climate change. Note that a higher level of BOD in industrial liquid effluents are major environmental and ecological issues. If these effluents are discharged into water bodies (pond, lake, river, etc.), the oxygen concentration in the water will decrease and aquatic species (water animals) would suffer and even die. If biological waste is converted into biogas, the BOD level in the effluent will decrease. Biogas can fulfill the energy demand (for cooking, agricultural engines (tractors), and power generator set) of rural villages as the villages have rich resources of biomass for production of biogas. The fuel properties of biogas and enriched biogas are given in Table 3.6.

TABLE 3.6

Physico-Chemical Properties of Biogas and Enriched Biogas

Properties	Units	Enriched Biogas	Raw Biogas
Composition	% (v/v)	CH_4 – 93%	CH_4 – 65%
		CO_2 – 4%	CO_2 – 33%
		H_2 – .06%	H_2 – .02%
		N_2 – 2.94%	N_2 – 1.98%
		H_2S – 20 ppm	H_2S – 500 ppm
Lower heating value	MJ/kg	42.62	20.5
Relative density	–	0.714	1.014
Flame speed	cm/s	–	25
Stoichiometric air-fuel ratio	–	17.16	17.16
Autoignition temperature	°C	–	650

Source: Subramanian, K. A., Mathad Vinaya, C., Vijay, V. K. and Subbarao, P. M. V. (2013), Comparative evaluation of emission and fuel economy of an automotive spark ignition vehicle fuelled with methane enriched biogas and CNG using chassis dynamometer, *Applied Energy*, 105, 17–29. (Reprinted with permission.)

A typical composition of raw biogas is methane (50%–60%), CO_2 (30%–40%), and traces of H_2S, N_2, and H_2 resulting to low calorific value. Even though raw biogas can be utilized in an automotive vehicle, the problem is that there is a drop in power and torque. OEMs will not accept adopting raw biogas as fuel for their vehicles. In addition, high CO_2 and H_2S in raw biogas would lead to problems such as gas storage in existing CNG cylinders, gas injecting into a natural gas grid, and gas transportation and distribution through existing natural gas pipelines. Hence, the fuel quality of raw biogas needs to be upgraded mainly by removing CO_2 gas and the upgraded biogas, called enriched biogas, can be used as a supplementary fuel in the transportation sector.

The calorific value of raw biogas is less compared to CNG fuel due to a higher CO_2 content. Further, the CO_2 content in raw biogas makes it difficult to store in cylinders due to embrittlement. These drawbacks of raw biogas necessitate enrichment in order to use it as an automotive fuel like CNG. A significant drop in brake power is expected for raw biogas compared to CNG due to less heat content.

3.9 Hydrogen

The molecular structure of hydrogen is

$$H ——————— H$$

Hydrogen (H_2)

Hydrogen is a carbon-free fuel. This is a highly reactive fuel. The important properties of hydrogen are given in Table 3.7.

As the density of hydrogen is low, a relatively large-sized fuel tank is needed to store it, which is one of the biggest technical issues. The flame velocity of hydrogen is the highest

TABLE 3.7

Properties of Hydrogen

Property	Units	Hydrogen
Molecular weight	g/mol	2.016
Density	kg/m³	0.0824
Flammability limits	Volume % in air	4–75
Auto-ignition temperature	K	858
Minimum ignition energy	mJ	0.02
Laminar flame speed	cm/s	185
Adiabatic flame temperature	K	2480
Quenching distance	mm	0.64
Lower heating value (MJ/kg)	–	119.7

Source: White, C. M., Steeper, R. R. and Lutz, A. E. (2006), The hydrogen-fueled internal combustion engine: A technical review, *International Journal of Hydrogen Energy*, 31(10), 1292–1305. (Reprinted with permission.)

compared to all hydrocarbon fuels. The combustion in heat engines fueled with hydrogen proceeds at a faster rate. The calorific value of hydrogen fuel in terms of mass is the highest but the volumetric heat content is lower, resulting in power drop. Hydrogen has a wider flammability range, which is advantageous because a hydrogen-fueled engine could operate on a very lean mixture. Hydrogen can also be blended with a slow-burning fuel like CNG to improve combustion performance. Hydrogen-fueled engines could eliminate carbon-based emissions (CO, HC, CO_2, PM).

3.10 Biodiesel

The physico-chemical properties of various biodiesels are given in Table 3.8.

Biodiesel is one of the main biofuels used for compression-ignition engines since biofuels (methanol, ethanol, butanol, biogas, hydrogen, producer gas, etc.) for spark-ignition engines are plentiful but biofuels for compression-ignition engines are relatively rarer. The applications of biofuels include transportation sectors, off-road transportation, construction machines, agricultural machines, and industrial power generation. The higher cetane number and oxygen content makes biodiesel a preferred fuel for compression-ignition engines by itself or as a blend with diesel. Biodiesel has no sulfur, aromatics, and other carcinogenic substances, and so biodiesel-fueled engines emit less emissions such as particulate matter, oxides of sulfur, and polycyclic aromatic hydrocarbons. Oxygen-embedded fuel leads to complete combustion. A high cetane number helps to reduce the cold-start ability problem, ignition delay, and engine knock. Due to a higher bulk modulus of biodiesel, in-line fuel-injection pressure and spray penetration are higher with all biodiesel blends. The spray-cone angle for biodiesel-diesel blends is lower than base diesel due to a higher density of biodiesel-diesel blends than base diesel. Ignition delay and rate of pressure rise decreased with all biodiesel blends due to a higher cetane number. The higher flash point of biodiesel over diesel allows the storage and handling of neat fuel or blends that are less complex. Poor cold-flow properties and oxidation stability are issues that need to be addressed with biodiesel.

3.11 F-T Diesel

The fuel properties of F-T diesel are given in Table 3.9.

F-T diesel is more of a paraffinic molecular structure that has almost no aromatics, olefins, or sulfur content. F-T diesel offers the benefit of a low carbon-to-hydrogen ratio compared to diesel fuel, which helps to reduce PM emission. F-T diesel is preferred over DME due to significant benefits in terms of storage in the automobile. F-T diesel has a higher cetane number, which would improve the transient performance of an engine. F-T diesel also has a higher calorific value in comparison with biodiesel, and so the specific power output of an F-T-diesel-fueled engine would improve. F-T diesel may require lubricity-improving additives to avoid wear of fuel-injection system components.

TABLE 3.8

Physico-Chemical Properties of Biodiesel

Property	Units	Diesel	ASTM Biodiesel D6751	First-Generation Biodiesel				Second-Generation Biodiesel			Third-Generation Biodiesel
				Palm	Linseed	Rapeseed	SOY	Karanja	Jatropha	Waste Palm Oil	Algae
Density	kg/m^3	837.8	–	878.4	865	884.9	885.2	860	897	875	872
Viscosity	mm^2/s	3.275	1.9–6.0	4.698	4.2	4.585	4.057	6.87	5.65	4.4	5.82
Cetane number	–	50.6	≥47	62	48	54.5	51.3	50	49	60.4	–
Calorific value	MJ/kg	45.85	–	39.91	40.759	39.9	39.66	38.5	37.9	38.73	40.8
Flash point	°C	68	130	189	161	177	173	175	187	70.6	–
Pour point	°C	–35	–	–	–18	–	–	–	–	–	–16
Distillation (50% volume)	°C	–	–	–	–	320	319	–	–	331	–
Distillation (90% volume)	°C	–	–	–	–	339	337	–	–	348	–
Sulfur	–	0.0019	–	0.0005	–	0.0001	0.0003	–	–	–	–

Source: Lin, B.-F., Huang, J.-H. and Huang, D.-Y. (2009), Experimental study of the effects of vegetable oil methyl ester on DI diesel engine performance characteristics and pollutant emissions, *Fuel*, 88(9), 1779–1785; Nautiyal, P., Subramanian, K. A. and Dastidar, M. G. (2014), Production and characterization of biodiesel from algae, *Fuel Processing Technology*, 120, 79–88.

TABLE 3.9

Fuel Properties of F-T Diesel

Property	Units	F-T Diesel
Density (at 15°C)	kg/m³	777.5
Viscosity (at 40°C)	mm²/s	2.547
Oxidation stability	g/m³	0.8
Lubricity (at 60°C)	mm	235
Cetane number	–	73.9
Low heating value	MJ/kg	43.53

Source: Mancaruso, E., Sequino, L. and Vaglieco Bianca M. (2011), First and second generation biodiesels spray characterization in a diesel engine, *Fuel*, 90(9), 2870–2883. (Reprinted with permission.)

3.12 DME

The molecular structure of DME is

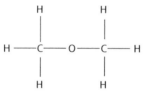

Dimethyl ether or methyloxymethane

The fuel properties of DME are given in Table 3.10.

TABLE 3.10

Physico-Chemical Properties of DME

Property	Units	DME
Chemical formula	–	$CH_3–O–CH_3$
Molar mass	g/mol	46
Density	kg/m³	667
Cetane number	–	>55
Autoignition temperature	K	508
Stoichiometric air/fuel ratio	–	9
Boiling point (at 1 atm)	K	248.1
Enthalpy of vaporization	kJ/kg	467.13
Lower heating value	MJ/kg	27.6
Kinematic viscosity of liquid	cS	<.1
Vapor pressure (at 298 K)	kPa	530

Source: Arcoumanis et al. (2008), The potential of di-methyl ether (DME) as an alternative fuel for compression-ignition engines: A review, *Fuel*, 87(7), 1014–1030. (Reprinted with permission.)

Like biodiesel, DME has a better cetane number and higher oxygen content, which makes it a better compression-ignition fuel. DME, which is liquid while stored, is a gaseous fuel that results in better diffusivity and air-fuel mixture formation. This allows the mixing to be much easier and hence result in better combustion. Particulate matter from diesel-fueled compression-ignition engines is a major problem and researchers face challenges to reduce the required levels. In some major cities, including New Delhi, selling diesel vehicles has been temporarily banned due to air pollution at an alarming stage. Note that solid-fueled thermal plants (coal-based power plants) emit the highest level of particulate matter whereas liquid-fueled compression-ignition engines emit moderate levels of PM emissions. However, even though PM emissions are less than 1% of fuel mass, the size of the particles and number of counts are high and therefore are not at an acceptable level. PM emissions are lower in gaseous-fueled engines than combustion with liquid fuel. In this respect, DME, which is a rare high-cetane-number fuel in a gaseous form, could enhance the sustainability of diesel-fueled engines and vehicles. DME is a preferred fuel over F-T diesel due to better mixing and combustion characteristics. The major drawback for DME compared to F-T diesel is that the calorific value is lower due to oxygen content, resulting in a power drop of the engine with DME fuel.

Summary

Biofuel quality for spark-ignition engines:

- The octane number of biofuels (methanol, ethanol, butanol, biogas, and hydrogen) is higher than that of petroleum-derived gasoline fuels.
- Olefins and aromatic-structured hydrocarbons are less in biofuels.
- The sulfur content in biofuels is generally lower or nil.
- The flash point is higher with biofuels, leading to better safety while handling the fuels
- Cold-flow characteristics (pour point, CFFP, etc.) are comparable with biofuels

Biofuel quality for compression-ignition engines:

- The cetane number of biofuels (biodiesel, F-T diesel, and DME) is higher than that of petroleum-derived diesel fuels.
- Olefins and aromatic-structured hydrocarbons are less in biofuels.
- The flash point is higher with biodiesel and F-T fuels than with base diesel; however, it is lower with DME fuel.
- Cold-flow properties are poor with biodiesel, whereas they are better with F-T diesel and DME fuels.

Overall, biofuel has better fuel quality than petroleum-derived fuels, and therefore is more preferable for use in engines. If biofuel quality is within the range of conventional/reference fuels (gasoline or diesel), the fuel can immediately be implemented in in-use vehicle fleets. Therefore, biorefineries are trying to maintain fuel quality with the reference fuel (conventional fuel). If the fuel quality is not the same and its properties vary too much from conventional fuel (e.g., hydrogen and DME are different from gasoline and diesel), the infrastructure of the entire fuel chain, including storage, transportation, and dispensing, has to be altered, modified, or newly built. This process would take a great deal of time, especially to implement the new fuel in vehicle fleets. If fuel quality is only slightly different from that of conventional fuel, vehicle technology can be tuned and calibrated without major modification for implementing biofuel. Therefore, fuel quality must be in line with vehicle technology.

References

Arcoumanis, C., Bae, C., Crookes, R. and Kinoshita, E. (2008), The potential of di-methyl ether (DME) as an alternative fuel for compression-ignition engines: A review, *Fuel*, 87(7), 1014–1030.

Bielaczyc, P., Woodburn, J., Klimkiewicz, D., Pajdowski, P. and Szczotka, A. (2013), An examination of the effect of ethanol–gasoline blends' physicochemical properties on emissions from a light-duty spark ignition engine, *Fuel Processing Technology*, 107, 50–63.

Lin, B.-F., Huang, J.-H. and Huang, D.-Y. (2009), Experimental study of the effects of vegetable oil methyl ester on DI diesel engine performance characteristics and pollutant emissions, *Fuel*, 88(9), 1779–1785.

Mancaruso, E., Sequino, L. and Vaglieco Bianca M. (2011), First and second generation biodiesels spray characterization in a diesel engine, *Fuel*, 90(9), 2870–2883.

Nautiyal, P., Subramanian, K. A. and Dastidar, M. G. (2014), Production and characterization of biodiesel from algae, *Fuel Processing Technology*, 120, 79–88.

Rutz, D. and Janssen, R. (2006), Project: Biofuel Marketplace (EIE/05/022/SI2.420009), *Overview and Recommendations on Biofuel Standards for Transport in the EU*, WIP Renewable Energies, Germany.

da Silva Trindade, W. R. and dos Santos, R. G. (2017), Review on the characteristics of butanol, its production and use as fuel in internal combustion engines, *Renewable and Sustainable Energy Reviews*, 69, 642–651.

Subramanian, K. A., Mathad Vinaya, C., Vijay, V. K. and Subbarao, P. M. V. (2013), Comparative evaluation of emission and fuel economy of an automotive spark ignition vehicle fuelled with methane enriched biogas and CNG using chassis dynamometer, *Applied Energy*, 105, 17–29.

The Gazette of India, Part II, Section 3, Subsection (i), Ministry of Road Transport and Highways Notification, September 16, 2016, New Delhi.

White, C. M., Steeper, R. R. and Lutz, A. E. (2006), The hydrogen-fueled internal combustion engine: A technical review, *International Journal of Hydrogen Energy*, 31(10), 1292–1305.

Zhen, X. and Wang, Y. (2015), An overview of methanol as an internal combustion engine fuel, *Renewable and Sustainable Energy Reviews*, 52, 477–493.

4

Introduction to Internal Combustion Engines

4.1 Introduction

Energy exists in several forms such as chemical, thermal, potential, kinetic, light, nuclear, magnetic, and electrical energy. Petroleum fuels such as gasoline, diesel, kerosene, propane, and butane have chemical energy that can be converted into heat and then work using heat engines. A heat engine is a device for producing motive power from heat.

An internal combustion engine, which is a heat engine, produces mechanical power by converting chemical energy into heat energy and then mechanical power. "Internal combustion" means combustion of fuels in a closed system (closed boundary system). In an internal combustion engine, heat is released by exothermic reaction of reactants (air and fuel) in a closed system when both intake and exhaust valves are closed. Heat energy is released by burning the air-fuel charge in the combustion chamber and then the heat energy is converted to mechanical power by movement of piston motion (movement of system boundary).

Heat and work are both transient phenomena and systems cannot possess heat or work. When a system undergoes a change, heat or work transfer may occur. Both heat and work cross the boundary of the system. Heat is defined as the form of energy that is transferred across the boundary by virtue of difference of temperature or temperature gradient. Work done is due to change in boundary of the system. The work done by the system and on the system are represented by + and − and heat flowing into the system and flowing out of the system are referenced as + and −.

Thermodynamics is a branch of physical science that deals with the relations between heat and other forms of energy. Thermodynamics deals with the conversion of chemical energy to heat and then work. Heat is defined as energy in transit. Work done is defined as the product of the force (combustion pressure X area) and the distance (piston movement) due to expansion of working fluid (boundary) by combustion. A thermodynamic system can be described using thermodynamics coordinates. The main thermodynamics coordinates are pressure (P), volume (V), temperature (T), and entropy (S) that are used for assessment of performance of a heat engine. The heat available in the system can be described using the coordinates of temperature and entropy whereas work done by the system is dealt by pressure and volume. P-V and T-S diagrams are used for assessment of heat and work availability in the system.

4.2 First Law of Thermodynamics

The first law of thermodynamics is based on the law of conservation of energy, which states that energy is neither created nor destroyed but can be transformed from one form to another. The heat given to the closed system is equal to summation of change in internal energy and work done, as shown in

$$\partial Q = dU + PdV \tag{4.1}$$

where
    Q = Heat given to the system (J)
    U = Internal energy (J)
    P = Pressure (N/m^2)
    V = Volume (m^3)

If combustion takes place at a constant volume with no work transfer, Equation 4.1 can be rewritten as given in

$$\partial Q = dU \tag{4.2}$$

Heat release rate can be determined with inputs of in-cylinder pressure and temperature, specific heat with respect to in-cylinder temperature, and instantaneous cylinder volume. Important combustion characteristics such as combustion duration and cumulative heat release can further be interpreted from the heat release rate diagram.

If a heat engine works with an isentropic compression and expansion process and an isothermal heat addition and rejection process, the efficiency of the heat engine is equal to Carnot efficiency. The Carnot efficiency depends on maximum temperature of the system and atmospheric or sink temperature. It can be interpreted through the first law of thermodynamics that the available heat in an engine can completely be converted to work. Any heat loss in the system is recoverable and reversible. It does not explicitly state how much heat loss, which is unutilized by the system, can be recovered and reutilized by the system or by another system. The direction of heat in the system is not also defined by the first law of thermodynamics. The second law of thermodynamics provides more clarity to these shortcomings.

4.3 Second Law of Thermodynamics

The actual heat engine does not provide the thermal efficiency equal to Carnot efficiency due to heat loss to the system and its surroundings. In all heat engines, there is always loss of heat in the form of conduction, radiation, and friction. The loss of heat/work in the system may be recovered back if they are reversible. A reversible process does not increase entropy (of the system and its surroundings) and the system is in thermodynamic equilibrium with its surroundings throughout the entire process. For example, if a system operates with source heat (H_1) with temperature (T_1), and sink heat (H_2)

with temperature (T_2), Equation 4.3 can be written for a complete reversible process. Otherwise, the entropy of the working substance in a reversible process can be written as shown in the equation.

$$\frac{H_1}{T_1} = \frac{H_2}{T_2}$$ (4.3)

The change in entropy of the working substance in a complete reversible process is zero, as shown in

$$\frac{H_1}{T_1} - \frac{H_2}{T_2} = 0$$ (4.4)

In general, a process that is not reversible is called an irreversible process. In a system, most of the heat is used for work transfer and some heat may be lost through heat transfer by conduction and radiation. The lost heat that can never be retrieved is called an irreversible process. For example, the heat lost by a hot body is ideally equal to the heat gained by a cold body. However, the cold body cannot gain the heat completely from the hot system due to loss of heat to the atmosphere due to conduction and radiation.

Loss of entropy of the hot body $= H/T_1$

Loss of entropy of the hot body $= H/T_2$

The total increase in entropy of the system is given in

$$\frac{H}{T_2} - \frac{H}{T_1}$$ (4.5)

It is a positive quantity as T_2 is less than T_1 due to the heat loss through conduction and radiation. Therefore, the entropy of the system increases in all irreversible processes. In all heat engines, there is always loss of heat in the form of conduction, radiation, and friction. Thus, change in entropy is not equal to zero but it is a positive quantity, as shown in

$$\frac{H_1}{T_1} - \frac{H_2}{T_2} \neq 0$$ (4.6)

The term entropy is important for assessing irreversibility in a thermal system. Entropy is a thermodynamic quantity representing the unavailability of a system's thermal energy for conversion into mechanical work and also a measure of the disorder of the molecules of the system. Molecular disorder increases when ice at solid phase transforms to liquid and vice versa. The entropy of gaseous state molecules is the highest compared to liquid and solid state due to gaseous molecules, which can freely move.

An internal combustion engine generates power by converting from chemical energy to heat during combustion. The power output of a spark-ignition engines and compression-ignition engines are in the range from few kilowatts (kW) to several megawatts (MW). Therefore, these engines are used in a variety of applications from agricultural pesticide sprayers (less than 1 kW) to two-wheeler and multiwheeler trucks (from approximately 5 kW to several MW) and electrical power generation (from approximately 2 kW to several MW). Spark-ignition engine is the highest specific power output compared to other heat engines as these engines can operate at an equivalence ratio of one. Based on type of ignition, an internal combustion engine can be classified into two categories:

1. Spark-ignition engine
2. Compression-ignition engine

A spark-ignition engine operates with the thermodynamic cycle of an Otto cycle, whereas a compression-ignition engine operates with a constant pressure cycle/dual cycle. The principle of operation of a spark-ignition engine and a compression-ignition engine is described next.

4.4 Otto Cycle

A spark-ignition engine operates with an Otto cycle. Air and fuel are inducted during suction stroke (0-1) and compresses the working fluid (1-2) during compression stroke. Ignition energy generated by a spark plug at end of the compression stroke initiates a chemical reaction of the charge and then combustion proceeds. Heat energy is released at a constant volume, resulting in an increase in in-cylinder pressure and temperature. The working fluid at high temperature and pressure pushes down the piston movement through which the heat energy is converted to mechanical work/power (3-4). The burned product (exhaust gas) during exhaust is expelled out to the atmosphere. Thus, a cycle is completed and then next cycle continues. The details of an Otto cycle process are given in Table 4.1. The pressure-volume and temperature-entropy diagram of an Otto cycle are shown in Figure 4.1.

TABLE 4.1

Details of Otto Cycle Processes

Process	Assumption
Suction (0-1)	Constant pressure
Compression (1-2)	Isentropic (adiabatic and reversible), $s_1 = s_2$
Heat addition (2-3)	(i) Adiabatic (ii) At constant volume (iii) Complete combustion
Expansion (3-4)	Isentropic (adiabatic and reversible), $s_3 = s_4$
Exhaust (1-0)	(i) Adiabatic (ii) Valves open and close at top and bottom center (iii) Process (4-1) at constant volume (iv) Process (1-0) at constant pressure

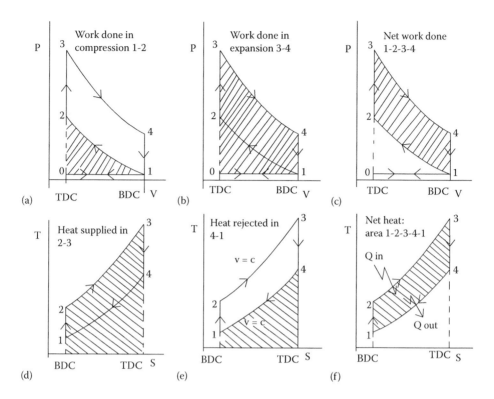

FIGURE 4.1
Pressure-volume diagram (a–c) and temperature-entropy diagram (d–f) for an Otto cycle based spark-ignition engine.

4.4.1 Derivation for Thermal Efficiency of Otto Cycle Based Spark-Ignition Engines

4.4.1.1 Process 1-2

Isentropic Compression: The working fluid (charge: air-fuel mixture) is compressed by a piston, which moves from bottom dead center (BDC) to top dead center (TDC). The pressure, temperature, and volume at BDC and TDC are denoted as P_1, T_1, V_1 and P_2, T_2, V_2, respectively. The specific heat ratio (γ) is defined by the ratio of specific heat at constant pressure (C_p) and specific heat at constant volume (C_v). The total volume of the engine's cylinder is summation of swept volume and clearance volume. V_1 and V_2 represent the engine's total volume and clearance volume, and the ratio of V_1 at BDC and V_2 at TDC is known as the compression ratio of internal combustion engines. While compressing the charge, in ideal condition, the change in entropy is constant and no heat transfer takes place from the inner cylinder to the outer surrounding atmosphere. This process is also called a reversible adiabatic compression process or isentropic compression process. The isentropic compression process is shown in

$$\frac{T_2}{T_1} = \left(\frac{p_2}{p_1}\right)^{\frac{\gamma-1}{\gamma}} = \left(\frac{v_1}{v_2}\right)^{\gamma-1} = r^{\gamma-1} \tag{4.7}$$

4.4.1.2 Process 2-3

Heat Addition at Constant Volume: The charge possessing chemical energy is released to heat energy during the combustion process. The chemical energy input to the engine is known as the heat addition at constant volume. During this process, the in-cylinder temperature and pressure rise sharply and are denoted as T_3 and P_3, respectively. The temperature T_3 is the equivalent of the adiabatic flame temperature of the chemical reaction. As the ratio of P_2/T_2 is equal to P_3/T_3 at constant volume combustion, the peak pressure or peak temperature can be found using the relation. The heat addition to the system can be calculated using

$$Q_a = m \times c_v \times (T_3 - T_2) \tag{4.8}$$

4.4.1.3 Process 3-4

Isentropic Expansion: The heat is converted to work during this isentropic expansion process. The combusted products (water vapor and carbon dioxide) with heat gained nitrogen gas at higher temperature and pressure pushes the piston toward the BDC position. In an ideal condition, the expansion of the combusted products with no heat transfer loss follows the isentropic process. The temperature, pressure, and volume at the TDC and BDC positions are called T_3, P_3, V_3 and T_4, P_4, V_4, respectively. The isentropic expansion process is shown in

$$\frac{T_3}{T_4} = \left(\frac{p_3}{p_4}\right)^{\frac{\gamma-1}{\gamma}} = \left(\frac{v_4}{v_3}\right)^{\gamma-1} = r^{\gamma-1} \tag{4.9}$$

4.4.1.4 Process 4-1

Heat Rejection at Constant Volume: The burned products are expelled from the in-cylinder to the atmosphere during this process. The pressure (P_4) of the exhaust gas is slightly higher than the atmospheric pressure. The heat rejection to the atmosphere can be calculated using

$$Q_r = m \times c_v \times (T_4 - T_1) \tag{4.10}$$

$$\text{Work done} = \text{Heat added} - \text{Heat rejected}$$
$$= mc_v(T_3 - T_2) - mc_v(T_4 - T_1)$$

$$\text{Thermal efficiency, } \eta = \frac{\text{Work done}}{\text{Heat supplied}}$$

$$\eta = \frac{mc_v(T_3 - T_2) - mc_v(T_4 - T_1)}{mc_v(T_3 - T_2)} \tag{4.11}$$

$$\eta = 1 - \frac{(T_4 - T_1)}{(T_3 - T_2)} \qquad (4.12)$$

Now, compression ratio $\dfrac{v_1}{v_2}$ = expansion ratio $\dfrac{v_4}{v_3}$ = r.

Also, $r = \dfrac{\text{Swept volume} + \text{Clearance volume}}{\text{Clearance volume}}$.

For ideal gas, $pv = RT$ and $pv^\gamma = $ constant

$$\frac{T_2}{T_1} = \left(\frac{v_1}{v_2}\right)^{\gamma-1} = \left(\frac{v_4}{v_3}\right)^{\gamma-1} = \frac{T_3}{T_4} = r^{\gamma-1}$$

$$\therefore \quad T_3 = T_4 r^{\gamma-1} \quad \text{and} \quad T_2 = T_1 r^{\gamma-1}$$

Hence, by substituting the above in Equation 4.12:

$$\eta = 1 - \frac{T_4 - T_1}{(T_4 - T_1)r^{\gamma-1}} = 1 - \frac{1}{r^{\gamma-1}}$$

$$\eta = \frac{1}{r^{\gamma-1}} \qquad (4.13)$$

Equation 4.13 shows that the thermal efficiency of an Otto cycle spark-ignition engine is the main function of compression ratio as it increases with an increasing compression ratio. Therefore, the compression ratio of a spark-ignition engine is increased from 5:1 to 11:1 as older engine was operated at 5:1 to 7:1 but engine is presently operated at 7:1 to 11:1 depending on the octane number of fuels. Compression ratio cannot be increased beyond the critical limit due to combustion with knock. Therefore, the thermal efficiency of the engine is limited.

Efficiency in terms of temperatures T_3 and T_4:

Since $\dfrac{v_1}{v_2} = \dfrac{v_3}{v_4}$, it follows that

$$\frac{T_2}{T_1} = \frac{T_3}{T_4} \quad \text{or} \quad \frac{T_2}{T_3} = \frac{T_1}{T_4}$$

and hence,

$$1 - \frac{T_2}{T_3} = 1 - \frac{T_1}{T_4}$$

or

$$\frac{T_3 - T_2}{T_3} = \frac{T_4 - T_1}{T_4}$$

Therefore,

$$\frac{T_4 - T_1}{T_3 - T_2} = \frac{T_4}{T_3}$$

Hence, substituting in Equation 4.12, we get

$$\eta = 1 - \frac{T_4}{T_3} = 1 - \frac{T_1}{T_2} \tag{4.14}$$

4.4.1.5 Mean Effective Pressure of Otto Cycle

Mean effective pressure is a measure of the effectiveness with which the displaced volume of the engine is used to produce work.

Let clearance volume be unity (i.e., $V_2 = V_3 = 1$ and $V_1 = V_4 = r$)

Let $\dfrac{p_3}{p_2} = \alpha$.

Now

$$\frac{p_2}{p_1} = \left(\frac{v_1}{v_2}\right)^\gamma = r^\gamma = \frac{p_3}{p_4}$$

$$\therefore \quad \frac{p_3}{p_2} = \frac{p_4}{p_1} = \alpha$$

Work done = area of the *p-v* diagram

$$
\begin{aligned}
&= \frac{p_3 v_3 - p_4 v_4}{\gamma - 1} = \frac{p_2 v_2 - p_1 v_1}{\gamma - 1} \\
&= \frac{1}{\gamma - 1}\left[p_4 r\left(\frac{p_3}{p_4 r} - 1\right) - p_1 r\left(\frac{p_2}{p_1 r} - 1\right) \right] \\
&= \frac{r}{\gamma - 1}\left[p_4(r^{\gamma-1} - 1) - p_1(r^{\gamma-1} - 1) \right] \\
&= \frac{r}{\gamma - 1}(r^{\gamma-1} - 1)(p_4 - p_1) \\
&= \frac{p_1 r(\alpha - 1)(r^{\gamma-1} - 1)}{\gamma - 1}
\end{aligned}
\tag{4.15}
$$

Length of the diagram $= r - 1$

$$\text{MEP} = \frac{\text{Area of the diagram}}{\text{Length of the diagram}}$$

$$\text{MEP} = \frac{p_1 r(\alpha - 1)(r^{\gamma-1} - 1)}{(\gamma - 1)(r - 1)} \quad (4.16)$$

4.5 Constant Pressure Cycle (Diesel Cycle)

A compression-ignition engine operates with a constant pressure cycle. Air from the atmosphere is only inducted during the suction stroke and then it is compressed during the compression stroke. Fuel (diesel or high cetane number fuel) is directly injected to the cylinder at the end of the compression stroke. The injected fuel is mixed with the surrounding air in the combustion chamber, and then a chemical reaction (ignition) of fuel with air proceeds. After the ignition process, combustion proceeds, resulting in conversion of chemical energy into heat energy. The heat addition is at a constant pressure known as a constant pressure cycle. The in-cylinder temperature and pressure increases and the hot burned gas pushes down the piston toward BDC; this process is called expansion. The heat energy is converted to mechanical work by piston movement (change in boundary of the system). The burned product is expelled to the atmosphere during the exhaust stroke and this cycle is completed, and then the next cycle continues in this manner. The details of the diesel cycle process are given in Table 4.2. The pressure-volume and temperature-entropy diagram of diesel cycle are shown in Figure 4.2.

4.5.1 Derivation of Thermal Efficiency and Mean Effective Pressure for Constant Pressure Cycle

4.5.1.1 Process 1-2

Isentropic compression:

$$\frac{T_2}{T_1} = \left(\frac{p_2}{p_1}\right)^{\frac{\gamma-1}{\gamma}} = \left(\frac{v_1}{v_2}\right)^{\gamma-1} = r^{\gamma-1} \quad (4.17)$$

TABLE 4.2

Process of Constant Pressure Cycle (Diesel Cycle)

Process	Details
Process 1-2	Isentropic compression of air
Process 2-3	Heat addition at constant pressure
Process 3-4	Isentropic expansion
Process 4-1	Heat rejection at constant volume

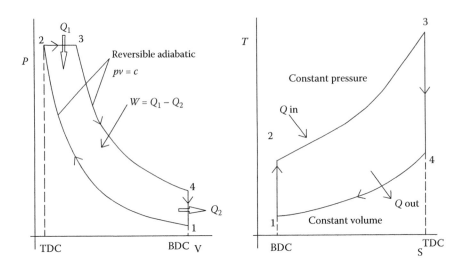

FIGURE 4.2
Pressure-volume and temperature-entropy diagram for diesel cycle.

4.5.1.2 Process 2-3

Heat addition at constant pressure:

$$Q_a = m \times c_p \times (T_3 - T_2) \tag{4.18}$$

4.5.1.3 Process 3-4

Isentropic expansion:

$$\frac{T_3}{T_4} = \left(\frac{p_3}{p_4}\right)^{\frac{\gamma-1}{\gamma}} = \left(\frac{v_4}{v_3}\right)^{\gamma-1} \tag{4.19}$$

4.5.1.4 Process 4-1

Heat rejection at constant volume:

$$Q_r = m \times c_v \times (T_4 - T_1) \tag{4.20}$$

The thermal efficiency of the ideal diesel cycle is given below:

$$\eta = \frac{\text{Heat added} - \text{Heat rejected}}{\text{Heat added}} = \frac{Q_1 - Q_2}{Q_1}$$

$$\eta = \frac{c_p(T_3 - T_2) - c_v(T_4 - T_1)}{c_p(T_3 - T_2)}$$

$$\eta = 1 - \frac{1}{\gamma}\left(\frac{T_4 - T_1}{T_3 - T_2}\right)$$

$$\eta = 1 - \frac{T_1}{\gamma T_2}\left(\frac{\frac{T_4}{T_1} - 1}{\frac{T_3}{T_2} - 1}\right) \tag{4.21}$$

For reversible adiabatic (isentropic) compression and expansion processes:

$$\frac{T_1}{T_2} = \left(\frac{v_2}{v_1}\right)^{\gamma - 1} \quad \text{and} \quad \frac{T_4}{T_3} = \left(\frac{v_3}{v_4}\right)^{\gamma - 1}$$

For constant pressure heat addition 2-3, $\frac{T_3}{T_2} = \frac{v_3}{v_2}$.
Also, $v_4 = v_1$.

Thus, $\dfrac{T_4}{T_1} = \dfrac{T_3}{T_2}\left(\dfrac{v_3/v_4}{v_2/v_1}\right)^{\gamma - 1} = \dfrac{v_3}{v_2}\left(\dfrac{v_3}{v_2}\right)^{\gamma - 1} = \left(\dfrac{v_3}{v_2}\right)^{\gamma}$

Substituting these values in Equation 4.21, the thermal efficiency can be written as

$$\eta = 1 - \frac{1}{\gamma\left(\dfrac{v_1}{v_2}\right)^{\gamma - 1}}\left[\frac{\left(\dfrac{v_3}{v_2}\right)^{\gamma} - 1}{\left(\dfrac{v_3}{v_2}\right) - 1}\right] \tag{4.22}$$

Substitution of cutoff ratio: v_3/v_2 in Equation 4.22: The ratio between volume (v_3) during constant pressure and clearance volume (v_2) is called the cutoff ratio, which is denoted by ρ.

$$\boxed{\eta = 1 - \frac{1}{r^{\gamma - 1}}\left[\frac{\rho^{\gamma} - 1}{\gamma(\rho - 1)}\right]} \tag{4.23}$$

Note that thermal efficiency increases with increasing compression ratio and also specific heat ratio, but decreases with increasing cutoff ratio.

4.5.1.5 Mean Effective Pressure of Diesel Cycle

Let the clearance volume be unity. Then

$$\text{Work done} = \text{Area of the } p\text{-}v \text{ diagram}$$

$$= p_2(v_3 - v_2) + \frac{p_3 v_3 - p_4 v_4}{\gamma - 1} - \frac{p_2 v_2 - p_1 v_1}{\gamma - 1}$$

$$= p_2(\rho - 1) + \frac{p_2 \rho - p_4 r - (p_2 - p_1 r)}{\gamma - 1}$$

$$= \frac{p_2(\rho - 1)(\gamma - 1) + p_2(\rho - \rho^\gamma r^{\gamma-1}) - p_2(1 - r^{1-\gamma})}{\gamma - 1}$$

$$= \frac{p_2}{\gamma - 1}\left[\gamma(\rho - 1) - r^{1-\gamma}(\rho^\gamma - 1)\right]$$

$$\text{MEP} = \frac{\text{Area of the indicator diagram}}{\text{Length of the indicator diagram}}$$

$$= \frac{p_2\left[\gamma(\rho - 1) - r^{1-\gamma}(\rho^\gamma - 1)\right]}{(\gamma - 1)(r - 1)}$$

$$\boxed{\text{MEP} = \frac{p_1 r^\gamma \left[\gamma(\rho - 1) - r^{1-\gamma}(\rho^\gamma - 1)\right]}{(\gamma - 1)(r - 1)}} \tag{4.24}$$

4.6 Dual Cycle

A dual cycle is similar to a constant pressure cycle except for the heat addition process. In a dual cycle, heat addition to the compression-ignition engine is by combined constant volume and constant pressure. The thermal efficiency of a heat engine is the highest when it operates with the addition of constant volume heat whereas the efficiency is less with the addition of constant pressure heat due to limitation of the maximum pressure. Hence, this cycle is also called a limited pressure cycle. All practical compression-ignition engines work with a dual cycle. The details of the dual cycle process are given in Table 4.3. Figure 4.3 shows a pressure–volume and temperature–entropy diagram of a dual cycle.

TABLE 4.3

Dual Cycle Process

Process	Details
Process 1-2	Isentropic compression of air
Process 2-3	Heat addition at constant volume
Process 3-4	Heat addition at constant pressure
Process 4-5	Isentropic expansion
Process 5-1	Heat rejection at constant volume

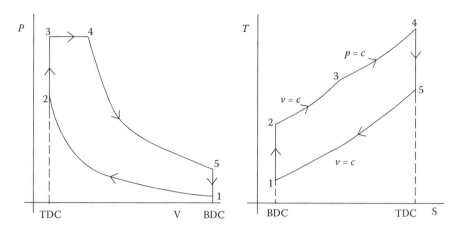

FIGURE 4.3
Pressure–volume diagram and temperature–entropy diagram for a dual cycle.

4.6.1 Derivation of Thermal Efficiency and Mean Effective Pressure

4.6.1.1 Process 1-2

Isentropic compression:

$$\frac{T_2}{T_1} = \left(\frac{p_2}{p_1}\right)^{\frac{\gamma-1}{\gamma}} = \left(\frac{v_1}{v_2}\right)^{\gamma-1} = r^{\gamma-1} \tag{4.25}$$

4.6.1.2 Process 2-3

Heat addition at constant volume:

$$Q_a = m \times c_v \times (T_3 - T_2) \tag{4.26}$$

4.6.1.3 Process 3-4

Heat addition at constant pressure:

$$Q_a = m \times c_p \times (T_4 - T_3) \tag{4.27}$$

4.6.1.4 Process 4-5

Isentropic expansion:

$$\frac{T_4}{T_5} = \left(\frac{p_4}{p_5}\right)^{\frac{\gamma-1}{\gamma}} = \left(\frac{v_5}{v_4}\right)^{\gamma-1} \tag{4.28}$$

4.6.1.5 Process 5-1

Heat rejection at constant volume:

$$Q_r = m \times c_v \times (T_5 - T_1) \tag{4.29}$$

The efficiency of this cycle can be written as shown in

$$\eta = \frac{\text{Heat supplied} - \text{Heat rejected}}{\text{Heat supplied}}$$

$$\eta = \frac{c_v(T_3 - T_2) + c_p(T_4 - T_3) - c_v(T_5 - T_1)}{c_v(T_3 - T_2) + c_p(T_4 - T_3)}$$

$$\eta = 1 - \frac{T_5 - 1}{(T_3 - T_2) + \gamma(T_4 - T_3)} \tag{4.30}$$

Now,

$$T_2 = T_1 \left(\frac{v_1}{v_2}\right)^{\gamma-1} = T_1 r^{\gamma-1},$$

$$T_3 = T_2 \frac{p_3}{p_2} = T_1 r^{\gamma-1} \cdot \alpha \left(\frac{v_1}{v_2} = r, \ \frac{p_3}{p_2} = \alpha\right)$$

$$T_4 = T_3 \frac{v_4}{v_3} = T_1 r^{\gamma-1} \cdot \alpha \cdot \rho \left(\frac{v_4}{v_3} = \rho\right)$$

$$T_5 = T_4 \left(\frac{v_4}{v_5}\right)^{\gamma-1} = T_1 r^{\gamma-1} \cdot \alpha \cdot \rho \left(\frac{v_4}{v_5}\right)^{\gamma-1}$$

Now,

$$\frac{v_4}{v_5} = \frac{v_4}{v_1} = \frac{v_4 v_3}{v_3 v_1} = \frac{v_4 v_2}{v_3 v_1}$$

since, $v_2 = v_3$

$$\therefore \ \frac{v_4}{v_5} = \frac{\rho}{r}$$

Hence,

$$T_5 = T_1 \alpha \rho^\gamma$$

Substituting for T_2, T_3, T_4, and T_5 into Equation 4.30, the efficiency can be written as

$$\eta = 1 - \frac{1}{r^{\gamma-1}}\left[\frac{\alpha\rho^{\gamma}-1}{(\alpha-1)+\alpha\gamma(\rho-1)}\right] \tag{4.31}$$

If cutoff ratio (ρ) = 1, it becomes an Otto cycle, and with α = 1, it becomes a diesel cycle. The efficiency increases with increasing compression ratio, α (pressure ratio of heat addition at constant volume) and specific heat ratio; however, it decreases with increasing cut of ratio.

Mean effective pressure of limited pressure cycle:

$$\text{Work done} = \text{Area of the } p\text{-}v \text{ diagram}$$

$$= p_3(v_4 - v_3) + \frac{p_4 v_4 - p_5 v_5}{\gamma - 1} - \frac{p_2 v_2 - p_1 v_1}{\gamma - 1}$$

$$= p_3 v_3(\rho-1) + \frac{(p_4 \rho v_3 - p_5 r v_3) - (p_2 v_3 - p_1 r v_3)}{\gamma - 1}$$

$$= \frac{p_3 v_3(\rho-1)(\gamma-1) + p_4 v_3\left(\rho - \dfrac{p_5}{p_4} r\right) - p_2 v_3\left(1 - \dfrac{p_1}{p_2} r\right)}{\gamma - 1}$$

Also,

$$\frac{p_5}{p_4} = \left(\frac{v_4}{v_5}\right)^{\gamma} = \left(\frac{\rho}{r}\right)^{\gamma} \quad \text{and} \quad \frac{p_2}{p_1} = \left(\frac{v_1}{v_2}\right)^{\gamma} = r^{\gamma}$$

Also,

$$p_3 = p_4, \; v_2 = v_3, \; v_5 = v_1$$

$$W = \frac{v_3\left[p_3(\rho-1)(\gamma-1) + p_3(\rho - \rho^{\gamma}\, r^{1-\gamma}) - p_2(1 - r^{1-\gamma})\right]}{\gamma - 1}$$

$$= \frac{p_2 v_2\left[\alpha(\rho-1)(\gamma-1) + \alpha(\rho - \rho^{\gamma}\, r^{1-\gamma}) - (1 - r^{1-\gamma})\right]}{\gamma - 1}$$

$$= \frac{p_1 r^{\gamma}\left(\dfrac{v_1}{r}\right)\left[\alpha\gamma(\rho-1) + (\alpha-1) - r^{1-\gamma}(\alpha\rho^{\gamma} - 1)\right]}{\gamma - 1}$$

$$= \frac{p_1 v_1 r^{\gamma-1}\left[\alpha\gamma(\rho-1) + (\alpha-1) - r^{1-\gamma}(\alpha\rho^{\gamma} - 1)\right]}{\gamma - 1}$$

$$\text{Mean effectiveness pressure } (p_m) = \frac{W}{v_1 - v_2} = \frac{W}{v_1\left(\dfrac{r-1}{r}\right)}$$

$$= \frac{p_1 v_1 r^{1-\gamma}\left[\alpha\gamma(\rho-1)+(\alpha-1)-r^{1-\gamma}(\alpha\rho^\gamma-1)\right]}{(\gamma-1)v_1\left(\dfrac{r-1}{r}\right)}$$

$$\text{MEP}(p_m) = \frac{p_1 r^\gamma\left[\alpha\gamma(\rho-1)+(\alpha-1)-r^{1-\gamma}(\alpha\rho^\gamma-1)\right]}{(\gamma-1)(r-1)} \qquad (4.32)$$

4.7 Two- and Four-Stroke Internal Combustion Engines

4.7.1 Two-Stroke Engine

A two-stroke engine is a type of internal combustion engine that completes a power cycle with every two strokes. One stroke means movement of a piston either upward or downward between TDC and BDC. Completion of a cycle in a two-stroke engine corresponds to one revolution of a flywheel (360-degree crank angle). Two-stroke engines have a higher specific power output than that of four-stroke engines, as well as relatively less numbers of moving parts and a lighter weight. The valve timing diagram of a two-stroke cycle engine is given in Figure 4.4.

A two-stroke engine has ports instead of valves. The lubricating oil mixed fuel is used for energy input as well as reducing friction between moving parts. Any two processes (compression + suction and expansion + suction) out of four happen simultaneously. The inducted charge is compressed when the piston moves downward during the expansion process, resulting in expansion of burned products and compression of inducted charge

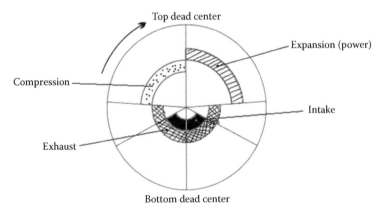

FIGURE 4.4
Valve timing diagram of a two-stroke cycle.

happening simultaneously. Similarly, when the piston moves toward TDC for compressing the charge, the charge for the next cycle is inducted into the crankcase. Hence, the suction and compression processes occur simultaneously. The exhaust gases are expelled using the fresh charge (air-fuel), which is called the scavenging process. The main disadvantage of a two-stroke engine is that some of the fresh charge escapes during the scavenging process, resulting in less thermal efficiency and higher levels of CO and HC emissions. Even though these engines were used in automobiles in the past because of their higher specific power output and compact engine system, today these engines are not used due to poor thermal efficiency and a higher level of emissions.

4.7.2 Four-Stroke Engine

A four-stroke engine is an IC engine in which the piston completes four separate strokes while turning a full rotation of the crankshaft. A stroke refers to movement of piston either traveling from TDC to BDC or vice versa. Four-stroke completions correspond to a 720-degree crank angle and completion of two revolutions of a flywheel. One stroke almost completely executes each process, such as suction, compression, expansion, and exhaust, and each stroke corresponding to each process is called a suction stroke, compression stroke, expansion stroke, and exhaust stroke. One cycle has four strokes and theoretically each stroke has a corresponding 180-degree crank angle rotation of a flywheel. A four-stroke engine has many advantages over a two-stroke engine in terms of higher thermal efficiency and less emissions, because sufficient time is relatively available for each process in a four-stroke engine compared to a two-stroke engine. The valve timing diagram of a four-stroke cycle engine is given in Figure 4.5.

The four strokes are explained below:

1. *Suction Stroke*: When the piston moves downward from TDC to BDC, the in-cylinder pressure is less than the atmospheric pressure, causing the charge (air-fuel) or air from the atmosphere to be sucked through the intake manifold. In this stroke, the intake valve is open until the end of the intake stroke and the exhaust

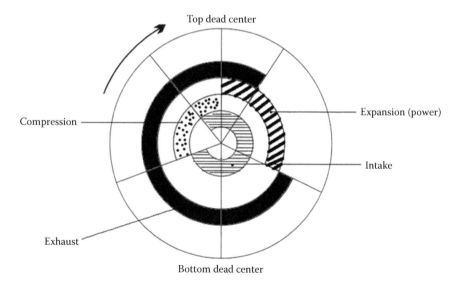

FIGURE 4.5
Valve timing diagram of a four-stroke cycle.

valve remains closed. This valve is closed when the piston touches BDC. Thus, a half rotation of the flywheel (0-degree to 180-degree crank angle) is completed.

2. *Compression Stroke*: The charge is compressed when the piston moves from BDC to TDC. The in-cylinder pressure and temperature increase during the compression stroke. Both valves are closed during the stroke. Another half rotation of the flywheel (181–360 degrees) is completed during this stroke.

3. *Expansion Stroke*: The spark, which is an external ignition energy generated by a spark plug, initiates ignition of the charge (air-fuel). The flame kernel grows and then exothermic heat release occurs by combustion. In the case of a compression-ignition engine, the high cetane number fuel at high pressure (300 to 2200 bar) is injected inside the cylinder. Heat transfer takes place from hot surrounding air to the fuel droplets. The self-ignition temperature of the fuel is lower than the surrounding compressed high-temperature air, resulting in autoignition of the fuel. The in-cylinder temperature and pressure increase due to combustion and thus the heat energy is converted into mechanical work by piston movement toward BDC. Both the intake and exhaust valves are closed during this stroke. Another half rotation of flywheel (361–540-degree crank angle) is completed during this stroke.

4. *Exhaust Stroke*: As the piston moves again toward TDC, the burned product is expelled to the atmosphere. Another half rotation of the flywheel (541–720-degree crank angle) is completed during this stroke. Thus, a cycle is completed and the next cycle continues from the suction stroke onward.

4.8 Classification of Internal Combustion Engines

Internal combustion engines can be classified in several ways:

1. Based on application:
   a. Automobile engine
   b. Aircraft engine
   c. Locomotive engine
   d. Marine engine
   e. Stationary engine
2. Based on ignition:
   a. Spark-ignition engine
   b. Compression-ignition engine
3. Based on engine cooling:
   a. Air-cooled engine
   b. Water (coolant: ethylene or propylene glycol and water) cooled engine
4. Based on basic engine design:
   a. Single cylinder, multicylinder inline, V, radial, opposed cylinder, opposed piston

5. Based on operating thermodynamic cycle:
   a. Atkinson (for complete expansion spark-ignition engine)
   b. Diesel (for the ideal diesel engine)
   c. Dual (for the actual diesel engine)
   d. Miller (for early/late inlet valve closing type spark-ignition (SI) engine)
   e. Otto (for the convectional SI engine)
6. Based on stroke:
   a. Four-stroke engine
   b. Two-stroke engine
7. Based on valve/port design and location:
   a. Design of valve/port
      i. Poppet valve
      ii. Rotatory valve
   b. Location of valve/port
      i. T-head
      ii. L-head
      iii. F-head
      iv. L-head
8. Based on fuel:
   a. Convectional fuel
      i. Crude oil derivatives: gasoline, diesel, propane, liquefied natural gas
   b. Alternative fuel (alternative to conventional fuels)
      i. Natural gas derived fuels: compressed natural gas
      ii. Coal derived fuels: methanol, DME
      iii. Biofuels and renewable fuels can be alternative fuels
   c. Biofuel
      i. Biomass-derived fuels: ethanol, butanol, biodiesel, biohydrogen, biogas, biodimethyl ether, bio-F-T diesel
   d. Renewable fuel
      i. Solar-hydrogen or renewable hydrogen produced through solar or wind energy system
9. Based on mixture preparation:
   a. Carburetion
   b. Fuel injection
10. Based on type of fuel injection system:
    a. Manifold injection
    b. Port injection
    c. Direct injection

11. Based on type of charge and ignition:

   a. Based on premixed/homogeneous charge with external ignition aid:

      Conventional Spark-Ignition Engine with Premixed/Homogeneous Charge: Because of external mixture preparation and more time availability during the suction stroke and compression stroke for mixing of air-fuel, almost half of the cycle time is available for air-fuel mixture preparation (i.e., 360 degree crank angle out of 720 degree crank angle per cycle). High octane number fuels such as gasoline, methane, propane, butane, biogas, producer gas, ethanol, methanol, butanol, and hydrogen have high self-ignition temperatures, and the fuel needs an external ignition source for initiating ignition, which is the spark ignition. A spark ignition engine is unable to ignite air-fuel charge itself using the heat available in the engine's combustion chamber during compression stroke. It may be noted that as the self-ignition temperature of high octane number fuels is higher than an engine's in-cylinder temperature, external ignition source to the engine is provided by a spark plug.

   b. Based on partially premixed and heterogeneous mixture with autoignition:

      Conventional Compression-Ignition Engine with Heterogeneous Mixture with Autoignition: The air-fuel mixture as a heterogeneous charge is prepared in a short fuel injection duration (total injection duration from approximately $10°$ to $50°$ crank angles out of $720°$ crank angles) between the end of the compression stroke and start of the expansion stroke, resulting in less time being available for mixture preparation and hence the mixture is heterogeneous. The fuels used for this type of engine are high cetane number fuels such as diesel, biodiesel, F-T diesel, and DME. A compression-ignition engine can ignite air-fuel mixture itself (called auto-ignition) using the heat (ignition source) available in the engine's combustion chamber at the end of compression stroke. In general, a high cetane number fuel needs less ignition energy than a high octane number fuel.

   c. Based on homogeneous charge with autoignition:

      i. *RCCI Engine*: The RCCI engine is a version of an HCCI engine. Both low-reactivity fuels (high octane number fuels), such as gasoline, ethanol, butanol, methanol, methane, biogas, and hydrogen, and high-reactivity fuels, such as diesel, biodiesel, F-T diesel, and DME, are used in an RCCI engine. A high reactivity fuel provides the required ignition energy to initiate ignition of a low reactivity fuel. The high reactivity fuel takes initially self-ignition and then it acts ignition source to ignite the low reactivity fuel. Thus, it can be concluded that ignition of a low reactivity fuel-air mixture is initiated by a high reactivity fuel, and hence no external ignition source by spark plug is necessary to initiate the ignition in an engine while it runs under the RCCI mode. For example, methane is inducted during a suction stroke and biodiesel is injected at the end of a compression stroke for initiation of ignition. Biodiesel (pilot fuel) receives ignition first as autoignition by the surrounding air-methane mixture and then this ignition energy is further used by the engine for igniting methane fuel. Methane has a high octane number and is suitable for spark-ignition engines but can be used in compression-ignition engines under RCCI mode.

      ii. *HCCI*: High octane number fuels are combusted by autoignition or without spark ignition. Similarly, the charge of high cetane number fuels is homogeneous instead of heterogeneous and is combusted by conventional autoignition. The drawback of spark-ignition engines is that flame combustion is disadvantageous in terms of better performance and emissions, but there are also advantages in terms of a premixed or homogeneous charge. In the case of compression-ignition engines, the main advantages include autoignition and that all charge is effectively burned, unlike flame combustion, but the drawback is a heterogeneous mixture. The advantages of both ignition engines are combined together into a hybrid called an HCCI engine. For example, methane is inducted or injected during a suction stroke and autoignition is initiated by a higher compression ratio of the engine or heating of charge. This combustion is also called low temperature combustion (LTC) because the localized temperature in a combustion chamber in a CI/SI engine is relatively lower due to a more homogeneous charge and elimination of flame combustion/combustion with heterogeneous mixture, resulting in an ultra-NOx emission.

12. Based on combustion chamber design:
    a. Open chamber: Disc, wedge, hemispherical, bowl-in-piston, bathtub
    b. Divided chamber:
       i.   (for CI): (1) Swirl chamber, (2) prechamber
       ii. (for SI): (1) Component vortex controlled combustion, (2) other designs

13. Based on type of air/charge induction:
    a. *Naturally Aspirated Engine*: The air/charge is inducted during a suction stroke based on pressure gradient between the in-cylinder and engine manifold. The inside pressure is less than atmosphere during a suction stroke, resulting in sucking the ambient air through the intake manifold.

    b. *Supercharged Engine*: An additional amount of air is supplied by a supercharger system that consists of a blower that takes power from the engine flywheel. The blower is generally connected with a flywheel pulley through the belt. The atmospheric air is sucked upstream of the blower and the compressed air is fed into intake manifold and gets mixed with the main inducted air. Thus, a greater amount of air is supplied to the engine and the engine can burn a relatively greater amount of fuel and deliver greater power output. The specific power output of the engine is higher than with a naturally aspirated engine. The power output can be increased using a supercharger without increasing swept volume. This supercharging system is highly useful when more torque is needed immediately, such as in emergency situations where a vehicle crosses into a flood stream, climbs over a steep hill, or passes over rough terrain, desert soil, an irregular stoned path, or pave-road.

    c. *Turbocharged Engine*: A turbocharged engine is similar to a supercharged engine except that the power for a functioning turbocharger is generated using waste exhaust gas. A turbocharger consists of a turbine-connected compressor and intercooler. The exhaust gas, which contains kinetic energy as well as thermal energy, is fed into the turbine, which converts the heat energy into mechanical energy (rotational), and the compressor, which is connected with the turbine, gets power from the turbine. The compressor sucks in the ambient air, the compressed air is cooled by an intercooler, and then the cooled compressed air is fed to the engine intake manifold. Both the main stream air and turbocharged air are mixed together in the engine's intake manifold and then entered into a cylinder during a suction stroke. Thus, an engine with a turbocharged system delivers higher power output than a naturally aspirated engine. The pumping work of the engine during a suction stroke is nearly zero with a turbocharger. Thermal efficiency of the engine is higher with a turbo-charger due to waste heat utilization.

14. Based on off-road vehicles:

Off-road vehicles are capable of running on both smooth and uneven surfaces. Applications of these engines or vehicles are in areas such as agriculture sectors (tractors, etc.) and prime movers for grinding, milling, and earth drilling for water well digging.

4.9 General Specifications of Internal Combustion Engines

The following terms and nomenclature associated with an engine are explained for better understanding of the working principle of IC engines (Figure 4.6). The rated power output

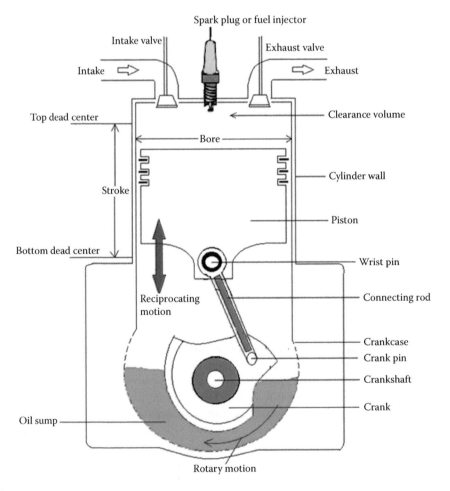

FIGURE 4.6
General specifications of an internal combustion engine.

and torque indicate that the engine can deliver the maximum power and torque at the specified speeds at a given condition. The specification of an engine includes its compression ratio, bore, stroke, number of cylinders, cylinder capacity (swept volume), rated power at corresponding speed, and type of fuel (single or multifuel compatibility). The compression ratio generally varies from 5:1 to 11:1 for spark-ignition engines and from 16:1 to 22:1 for compression-ignition engines.

The rated power is defined by the manufacturer of the engine (OEM) and an engine/vehicle at the rated power can develop maximum power. The engine's speed with rated torque is not the same as with rated power because maximum torque occurs at relatively lower speed compared to maximum power output. Torque is defined as the capability of doing work whereas power is defined as rate of work done. An engine or vehicle needs more torque as it starts to move from rest/idle condition. If the vehicle has to run on a road with unforeseen conditions such as flooding, loose soil, steep slope, or any other road obstacle, it needs relatively more torque. The torque-speed characteristics indicate that it increases with an increase in the engine's speed but decreases with a further increase in speed. Unlike torque, power output increases nonlinearly with an increase in the engine's speed. The torque-speed characteristics of an engine are important when choosing a vehicle for transportation of goods, and the recommended fuel and lubricants should be used for better engine life.

4.10 Automotive Vehicles

Automotive vehicles are mostly used for mass (goods) and passenger transportation. The speed, power, and torque of the engine/vehicle may vary based on its payload and traffic conditions.

Classification of automobiles are based on the number of wheels:

- Two-wheeler: motorcycles, scooters
- Three-wheeler: tempo, auto rickshaw
- Four-wheeler: car, jeep, bus, truck
- Six-wheeler: buses and trucks have six tires, four of which are carried on the rear wheels for additional traction
- Eight- or more wheeler truck (example: a heavy hauler, an armored vehicle)

4.11 Power Generator Set Engines with Alternator

An engine is coupled with a alternator that is called a generator or power genset engines. The constant engine speed is generally maintained as the alternator has to supply the electrical power with the desired frequency. The engine speed is generally either 1500/1800 rpm or 3000/3600 rpm for an electrical power supply with 50/60 Hz. The relation between engine or alternator speed and frequency is given in

$$N = \frac{120 \times f}{P} \tag{4.33}$$

where
  N = Speed of engine or alternator (rpm)
  F = Frequency (hertz)
  P = Number of alternator poles

The power output of internal combustion engines can be calculated using

$$BP = \frac{m_f \times CV \times \eta_t}{1000} \tag{4.34}$$

or Equation 4.34 can be written with a mass flow rate of air, as shown in

$$BP = \frac{m_a \times \varnothing \times CV \times \eta_t}{AFR_S \times 1000} \tag{4.35}$$

where
  BP = Brake power (kW)
  m_f = Mass flow rate of fuel (kg/s)
  m_a = Mass flow rate of air (kg/s)
  CV = Calorific value of fuel (kJ/kg)
  η_t = Thermal efficiency of engine
  AFR_s = Stoichiometric air-fuel ratio
  Φ = Equivalence ratio (<= 1)

4.12 Power Output and Emissions

Power output could increase with an increase in the mass flow rate of fuel, thermal efficiency, or both. If the fuel flow rate increases, the air flow rate has to be increased relatively; otherwise, combustion may be incomplete. If thermal efficiency and power output are constant, the mass flow rate of a fuel will decrease with an increase in the calorific value of the fuel. Brake power can be calculated using

$$BP = \frac{BMEP \times V_S \times N \times K \times 100}{NS \times 60} \tag{4.36}$$

where
  BP = Brake power
  $BMEP$ = Brake mean effective pressure (bar)
  V_s = Swept volume (m^3)
  N = Speed (rpm)
  NS = Number of strokes: 1 for two-stroke engine/2 for four-stroke engine
  K = Number of cylinders

Equation 4.36 indicates that brake power is a function of BMEP, swept volume, speed, and number of cylinders. BMEP, engine speed, and swept volume cannot be increased beyond a certain level. Therefore, the number of cylinders has to be increased in order to get more power output.

High octane number fuels such as gasoline, methanol, ethanol, butanol, natural gas, propane, butane, and hydrogen are suitable for spark-ignition engines whereas high cetane number fuels such as diesel, biodiesel, F-T diesel, and DME are suitable for compression-ignition engines.

The stoichiometric chemical reaction is shown in Equation 4.37. One mole of hydrocarbon burning with stoichiometric air quantity gives b mole of CO_2 and C mole of H_2O. Nitrogen does not affect the reaction and leaves as it is. Complete or perfect combustion means that the products are only carbon dioxide and water vapor and the combustion efficiency is 100%.

$$C_xH_y + a(O_2 + 3.76\,N_2) \rightarrow b\,CO_2 + c\,H_2O + a \times 3.76 \times N_2 \qquad (4.37)$$

Note that the combustion efficiency of a practical engine is less than 100%. The intermediate products will present in the exhaust gas if combustion ends before completion due to a variety of reasons. This means the intermediate products are unable to oxidize into carbon dioxide and water vapor. Products with incomplete combustion, which is called emissions, are given below:

- Carbon monoxide (CO)
- Hydrocarbon (HC)
- Carbon dioxide (CO_2)
- Water vapor (H_2O)
- Methane (CH_4)
- Hydrogen (H_2)
- Particulate matter (PM)
- Soot
- Smoke
- Nitrous Oxide (N_2O) N_2O*
- Oxides of Nitrogen (NOx): combined nitric oxide (NO) and nitrogen dioxide (NO_2) NOx* (NO and NO_2)

*N_2O and NOx do not depend on whether the combustion is complete or incomplete. These emissions mainly depend on in-cylinder temperature, availability of oxygen concentration in combustion chamber, residence time, and intermediate radicals.

4.13 Comparison of Internal Combustion Engines with Gas Turbine and Steam Turbine Engines

If heat engines (internal combustion engines, gas turbine, and steam turbine) are compared per kilowatt, the initial cost of internal combustion engine is the lowest compared to gas and steam turbines. A comparison between these heat engines are shown in Table 4.4.

TABLE 4.4

Comparison of Performance and Emissions of Heat Engines

Description (per kW)	Internal Combustion Engines (Diesel-/Gasoline-Fueled Engine)	Open-Cycle Gas Turbines (Natural Gas Based Power Plant)	Steam Turbine Engines (Coal-Based Power Plant)
Specific power output (kW/unit size) (swept volume/combustion chamber volume)	Highest	Moderate	Lowest
Torque	Highest	Moderate	Lowest
Speed	Lowest	Highest	Moderate
Fuel quality requirement	High cetane number/octane number fuels with ultrasulfur and no inorganic compounds	Fuels should be sulfur-free and have negligible inorganic compounds	Any type of fuel can be used but is currently limited in order to meet emission norms/regulations
Fuels	High octane number fuels for SI engines: Gasoline, methane, propane, butane, producer gas, biogas, ethanol, methanol, propanol, hydrogen High cetane number fuels for CI engines: Diesel, biodiesel, F-T diesel, DME	Fuels: Methane, propane, butane, and any liquid fuels Aircraft: Aviation kerosene (aviation turbine fuel [ATFl])	Fuels: Coal, biomass, industrial wastes (wood pulp, bagasse, etc.)
Thermal efficiency	Almost the same (35% to 39%) but it depends on operating condition	Almost the same (maximum from 35% to 41%)	Almost the same (maximum from 35% to 41%)
Frictional power	Highest	Much lower (only bearing and accessories)	Much lower (only bearing and accessories)
Air-fuel ratio	Stoichiometric air-fuel ratio	Lean air-fuel ratio	Lean
Maximum temperature range	Highest (>1800 to <2200 K), almost adiabatic flame temperature	Temperature is limited to 1500°C due to thermal stress limitation of turbine blades	Temperature is limited due to use of low fuel quality (coal, biomass, etc.) and poor mixing of air-fuel and poor combustion efficiency
Emissions: (CO, HC, PM, NOx)	PM and NOx: Higher for CI engine NOx: Higher for SI engine CO and HC: Moderate for both engines Note: PM is less if gaseous fuel is used. In this case, diesel (liquid fuel) combustion in CI engine contributes more PM emission	HC, CO, PM, and NOx: Lowest Note: PM emission is low due to gaseous fuel usage and is low if fuel is in a gaseous state due to better mixing of fuel with air	HC, CO, PM, and NOx: Highest Note: PM emission is the highest due to solid state fuel usage because solid state fuel is unable to mix with air, and also inorganic substance (ash) contributes more PM emission
Water requirement	Much lower	Lowest	Highest

(Continued)

TABLE 4.4 (CONTINUED)

Comparison of Performance and Emissions of Heat Engines

Description (per kW)	Internal Combustion Engines (Diesel-/Gasoline-Fueled Engine)	Open-Cycle Gas Turbines (Natural Gas Based Power Plant)	Steam Turbine Engines (Coal-Based Power Plant)
Ash	Nil	Nil	Highest
Water pollution	Almost nil	Almost nil	Highest
Land pollution	Nil	Nil	Highest (ash and other residues stored on land)
Air pollution	Moderate	Lowest	Highest
Noise pollution	Lowest	Highest	Highest
Space requirement	Lowest	Moderate	Highest
Personnel requirement	Much lower	Moderate	Highest
Time taken for starting of the engine/system	Much lower	Moderate	Highest
Success rate and failure (sabotaging) rate	Success rate: >98% Failure rate: <2% More reliable	Success rate: >98% Failure rate: <2% More reliable	Success rate: >98% Failure rate: <2% More reliable
Distribution of load/distribution of rated power (MW)	Best (e.g., 100 MW can be generated using 50 units of IC engine with each 2 MW power output, the power can vary from 2 to 100 MW)	Not flexible (2%, 5%, or 10% of rated power output are not possible)	Not flexible (2%, 5%, 10% of rated power output are not possible)
Continuous or discrete power generation/application	Discrete operation is preferable	Moderate	Moderate
Carbon capture storage (CCS) system	Not possible for automobiles but it is possible for stationary engines	Possible	Possible
Initial cost per kW	Lowest	High	High
Fuel cost ($/MJ)	High	High	Lowest
Operating cost ($/kW-hr or $/km)	Highest	High	Lowest
Cost of spare parts (piston, ring, cylinder heat, spark plug, injector, turbine blade, etc.)	Lowest	Highest	Highest
Cost for maintenance and service of engine	High	Moderate	Moderate
Carbon tax	High	Low	Highest

The total cost comparison between the three types of heat engines is given below:

Total cost of the system = Initial system (engine) cost + Interest cost
+ Operating cost (fuel and lubricants) + Maintenance cost
+ Personnel cost + Space cost + Carbon tax + Other cost

4.14 Selection of Heat Engines for Industrial Applications

Maintenance cost + Personnel cost + Space cost + Carbon tax + Other cost

The selection of heat engines based on the following conditions:

Discrete power generation from a few kilowatts to a few megawatts range for few hours per day: Internal combustion engines.

If an industry needs more electrical power (MW range) demand for a longer time duration (more than 16 hours) and the industry has continuous gaseous fuel availability, a GT power plant can be chosen.

If an industry needs combined electrical power and heat (in the form of steam) for a longer time duration (more than 16 hours) per day and the industry has resources and infrastructure for solid fuel, an ST power plant can be chosen.

If an industry needs continuous power supply, fuel availability would determine the suitability of a GT system or ST system. For example, India and China have ample reserves of coal fuel, so the ST system would be preferable for power generation. In the cases of middle Asia and Russia, where natural gas is more available, the GT system would be preferable for power generation. Since unit power cost is approximately equal to three fourths of fuel cost, fuel is a prime factor for selecting the type of power plant. Many industries have combined cycle heat engines in power plants for electrical power generation; for example, GT system + IC engine, GT system + ST system (combined cycle power plant), ST system + IC engine, and GT system + GT system + IC engine. Even though a GT or ST system alone can supply the desired power supply demand, IC engines are also additionally used for electrical power generation for meeting peak power demand, starting of the system, and emergency and back-up power reserves.

Selecting heat engines in industries is a complex procedure as many parameters need to be considered for specific industries. Further discussion of this topic is out of the scope of relevance for this book; however, some information will now be presented and discussed for a better understanding of IC engines in industrial applications.

Even though the operating and maintenance costs of an internal combustion engine are the highest, it is still a preferable choice because the initial cost of the system per kilowatt is much less compared to other engines. This comparison is only valid for power generation. In the case of transportation, an IC engine is a unique choice. Alternative vehicles such as fuel-cell and battery-operated vehicles are used for transportation but the number of existing vehicles is lower than the number of IC engine vehicles.

4.15 Comparison between Spark-Ignition Engines and Compression-Ignition Engines

A common question is frequently asked whether a spark-ignition engine could give better performance than a compression-ignition engine. Both engines can deliver the desired power output. The cost of a spark-ignition engine is less than that of a compression-ignition engine because the cost of a fuel injection system and the strength of material for withstanding a higher compression ratio in a compression-ignition engine is higher. A compression-ignition engine needs relatively frequent maintenance mainly due to servicing of the injection system. The deciding factor for choosing of the type of ignition engine is based on fuel cost. In India, the price of diesel is less than that of gasoline because diesel fuel is subsidized by the government for controlling the cost of mass/goods transportation. Diesel vehicles are mostly used for mass transportation whereas spark-ignition engines are used for passenger transportation. However, a diesel engine emits a high level of particulate matter. The existing technologies for diesel engines are unable to reduce the particle size and number to the desired level. Advanced technologies, including RCCI and HCCI, could reduce this emission at the source level. The comparison of spark-ignition engines and compression-ignition engines per kilowatt with the same swept volume is summarized in Table 4.5. The Carnot efficiency of a spark-ignition engine with a stoichiometric air-fuel ratio is high because maximum temperature is high, whereas thermal efficiency is high with a higher compression ratio in compression-ignition engines. A compression-ignition engine's swept volume is higher than that in a spark-ignition engine due to lean burn operation and less speed, resulting in an increase in frictional power. But theoretically, the thermal efficiency of a compression-ignition engine is higher than in an Otto cycle spark-ignition engine. Due to a steep increase in octane number of fuels, the compression ratio of a spark-ignition engine is increased, resulting in relatively better thermal efficiency than in previous engine models. Modern diesel engines operate with a fuel penalty because additional energy input is required for aftertreatment devices, such as diesel oxidation catalyst (DOC), particulate trap, and selective catalyst reduction (SCR), to function and also for higher injection pressure (>2000 bar). The advantage of high thermal efficiency with higher compression ratio may disappear by the fuel penalty due to the aftertreatment devices. Table 4.5 shows that if the fuel price is same for both engines, the net cost of a spark-ignition engine could be less than that of a compression-ignition engine.

An internal combustion engine emits emissions, including CO, HC, NOx, PM, CH_4, N_2O, and CO_2. Emissions such as CO and HC are reduced significantly by optimizing the engine's design and operating parameters and also with the help of aftertreatment devices. Therefore, modern engines still have the problem of NOx and particulate matter. Some important techniques and technologies are highlighted next.

TABLE 4.5

Comparison of Spark-Ignition Engines and Compression-Ignition Engines per Kilowatt with the Same Swept Volume

Description	Spark-Ignition Engines	Compression-Ignition Engines
Specific power output	Higher	Lower
Volumetric efficiency	Lower	Higher
Torque	Slightly lower due to less volumetric efficiency because of a throttle system	Higher as they do not have a throttle system
Mixture preparation	External (carburetor and manifold injection, port injection) and internal during compression stroke Only internal: Inside engine cylinder	Inside engine cylinder
Engine speed	Higher	Lower
Power output	Higher (even though torque is low, engine speed is high, resulting in higher net power output)	Lower
Compression ratio	Lower (generally from 5:1 to 11:1)	Higher (generally from 16:1 to 22:1)
Thermal efficiency	Lower, mainly due to lower compression ratio	Higher due to mainly higher compression ratio
Type of ignition	Spark ignition	Autoignition
Distinguishing components	Spark plug, induction/injection system, throttle system, quality governing	Injector and injection system, quantity governing
Air/charge motion	Charge motion: Tumble (organized rotation of charge perpendicular toward the cylinder axis)	Air motion: Swirl (organized rotation of charge toward the cylinder axis) and squish (radial inward motion of air at the end of a compression stroke)
Charge preparation time	Higher (almost half [1/2] of a cycle time for carburetor and manifold injection and almost one-fourth [1/4] of a cycle time for direct in-cylinder injection)	Much lower (almost one-fifteenth (1/15) of a cycle time)
Antiknocking property/ignition quality	High octane number (>85)	High cetane number (>45)
Self-ignition temperature of fuel	Higher	Lower
Mixing of air-fuel	Higher	Lower
Type of charge	Premixed charge	Premixed and heterogeneous charge
Fuels	Conventional fuels: Gasoline, methane, propane, butane Biofuels: Methanol, ethanol, butanol, biogas, hydrogen	Conventional fuels: Diesel, heavy fuel oil (HFO) Biofuels: Biodiesel, F-T diesel, DME
Air-fuel ratio	Stoichiometric	Lean due to smoke-limiting factor
CO and HC	Moderate	Lower or negligible due to lean mixture
NOx emission	Uncertain; dependent on operating conditions	Uncertain; dependent on operating conditions
Soot and smoke	Almost nil	Higher
Particulate matter	Lower	Higher
Weight of engine/kW	Lower	Higher
Volume of engine/kW	Lower	Higher

(Continued)

TABLE 4.5 (CONTINUED)

Comparison of Spark-Ignition Engines and Compression-Ignition Engines per Kilowatt with the Same Swept Volume

Description	Spark-Ignition Engines	Compression-Ignition Engines
Noise	Lower	Higher
Initial cost	Lower due to lower cost of lower injection pressure system (1 to 4 bar) Lower material cost due to lower compression ratio of engine	Higher due to higher cost of higher injection pressure system (400 to 2200 bar) High cost for material in order to withstand high compression ratio of engine
Maintenance and service cost	Lower	Higher due to fuel injection system
Operating cost	Dependent on fuel cost	Dependent on fuel cost
Net system cost if fuel cost is the same for both engines	Could be lower as compression ratio of SI engine increases due to a steep increase in octane number of fuels by refineries	Higher due to fuel penalty by additional energy input requirement to aftertreatment devices, such as diesel oxidation catalyst (DOC), particulate trap, and selective catalyst reduction (SCR), and also higher injection pressure (>2000 bar)

4.16 NOx Emission Reduction Techniques

4.16.1 EGR

EGR is a technique to reduce NOx at the source level in spark-ignition engines and compression-ignition engines. EGR can also be used to control autoignition in an HCCI engine. The trapped exhaust gas contains certain radicals that would be helpful to oxidize the intermediate products during combustion. Some quantity of exhaust gas that is trapped from the exhaust manifold is recycled to an engine cylinder through the engine's intake manifold, as shown in Figure 4.7.

EGR can be classified into two categories: hot EGR and cold EGR. Hot EGR is directly recirculated to an engine. The trapped exhaust gas from the engine's exhaust manifold is cooled by an intercooler and then the gas is supplied to the engine, which is called cold EGR. Hot EGR contains primarily carbon dioxide, water vapor, and nitrogen with traces of oxygen (if a lean burn engine), carbon monoxide, hydrocarbon, and oxides of nitrogen. The composition of cold EGR is similar to hot EGR except for the presence of water vapor and other condensable matter. The trapped exhaust gas is mixed along with the inducted air during suction stroke and then enters into the engine's cylinder. The stoichiometric equation for air without EGR, air with hot EGR, and air with cold EGR are shown in Equations 4.38 to 4.40.

Stoichiometric equation for a conventional engine:

$$C_xH_y + a(O_2 + 3.76\,N_2) \rightarrow bCO_2 + cH_2O$$
$$+ 3.76\,aN_2 + dCO + eHC + fNO + gNO_2 \tag{4.38}$$

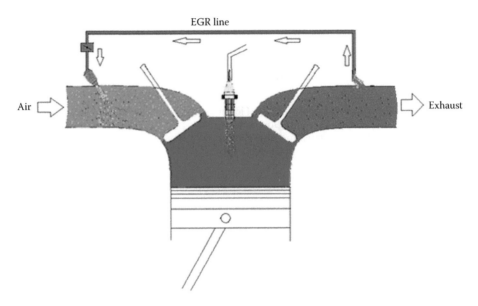

FIGURE 4.7
Schematic diagram of exhaust gas recirculation.

Stoichiometric equation for an engine with hot EGR:

$$
\begin{aligned}
C_xH_y + a(O_2 + 3.76\,N_2) &\rightarrow b_1CO_2 + c_1H_2O \\
&+ 3.76\,a_1N_2 + d_1CO + e_1HC + f_1NO + g_1NO_2 \rightarrow b_2CO_2 + c_2H_2O \\
&+ 3.76\,a_2N_2 + d_2CO + e_2HC + f_2NO + g_2NO_2
\end{aligned}
\tag{4.39}
$$

Stoichiometric equation for an engine with cold EGR:

$$
\begin{aligned}
C_xH_y + a(O_2 + 3.76\,N_2) &+ b_1CO_2 + 3.76\,a_1N_2 \\
&+ d_1CO + e_1HC + f_1NO + g_1NO_2 \rightarrow b_2CO_2 + c_2H_2O \\
&+ 3.76\,a_2N_2 + d_2CO + e_2HC + f_2NO + g_2NO_2
\end{aligned}
\tag{4.40}
$$

The EGR gas in an engine reduces oxides of nitrogen emission due to the following actions:

- Dilution of oxygen concentration due to recycled combustion products, which acts as a barrier between the nitrogen and oxygen
- Reduction of oxygen concentration as some of the oxygen/air is replaced by the circulated gas
- Decrease in in-cylinder temperature due to EGR acting as a heat sink mainly by CO_2 and N_2 (enriched), which increases specific heat of charge
- Less adiabatic flame temperature of the air-fuel mixture with EGR compared to the mixture without EGR

The EGR may negatively affect the performance of an engine due to the following:

- Drop in volumetric efficiency and power with a higher percentage of EGR and volumetric efficiency of cooled EGR is better than hot EGR
- Poor transient performance during start-up and warm-up periods
- Difficulty in maintaining a constant flow rate and composition while operating the engine at fluctuating loads
- Increase in PM/smoke emissions due to reduction in oxygen concentration
- Lower thermal efficiency due to slow combustion reaction
- Increased complexity of the engine with an EGR system

EGR slows down the ignition process, which is a disadvantage for conventional engines, but an advantage for controlling the start of combustion and autoignition in an HCCI engine. Cooled EGR could reduce cylinder pressure oscillations in an HCCI engine (Lin et al., 2016). A dedicated EGR (D-EGR), which was developed with reformatted gas (CO and hydrogen) produced using a dedicated cylinder (with rich mixtures), enables improved fuel economy, reduced emissions, combustion stability, and knock residence of engines (Robertson, 2017).

Higher HC emission at part load is a problem in an RCCI and this emission can be controlled using the optimum EGR due to increasing fuel-air ratio by some replacement of air and certain active radicals (Kalsi and Subramanian, 2016). Some earlier technical issues with EGR have been resolved by researchers, so EGR is now well established and can potentially be an effective method to reduce NOx emission at the source level along with other benefits. Therefore, EGR is almost mandatory for advanced engines to reduce NOx emission.

4.16.2 SCR

SCR is an aftertreatment method used to control NOx emission in an engine exhaust system. A liquid reductant (e.g., ammonia, urea solution, cyanuric acid, ammonia-blended sodium carbonate, and xenon lamp based reduction) is injected into the catalyst coated exhaust system in which NO is converted to N_2 and water. The SCR system consists of an SCR tank (for chemical solution), dosing injector, electronic control unit (ECU), and SCR catalyst. The reductant is injected into the hot exhaust gas trapped temporarily by the catalyst surface and the injector is controlled using the ECU.

Equations 4.41 and 4.42 show the ammonia reaction with NO and NO_2 and through these reactions, NO and NO_2 are converted to nitrogen and water vapor. These reactions may proceed in temperatures from 800°C to 1200°C without a catalyst and 260°C to 500°C with a catalyst.

$$4NO + 4NH_3 + O_2\text{-catalyst} \rightarrow 4N_2 + 6H_2O + \text{Heat} \qquad (4.41)$$

$$2NO_2 + 4NH_3 + O_2\text{-catalyst} \rightarrow 3N_2 + 6H_2O + \text{Heat} \qquad (4.42)$$

NOx conversion efficiency depends on many parameters, including temperature of exhaust gas (Kuihua and Chunmei, 2007), residence time, degree of homogeneous mixing of reductant with exhaust gas (Røjel et al., 2000), NO and NO_2 levels, and CO and O_2 concentration.

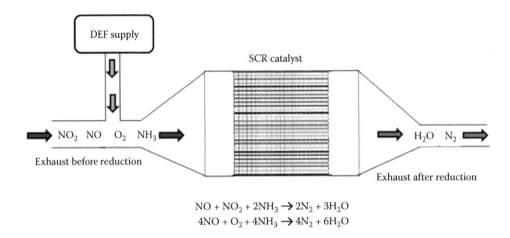

$$NO + NO_2 + 2NH_3 \rightarrow 2N_2 + 3H_2O$$
$$4NO + O_2 + 4NH_3 \rightarrow 4N_2 + 6H_2O$$

FIGURE 4.8
Schematic diagram of SCR.

Ammonia breaks down to NH_2 by O and OH radicals and then NH_2 is converted to nitrogen and water vapor. The HNCO is decomposed from cyanuric acid and converted to NCO and NH_2, which are NO reducing agents. In the case of urea, it is decomposed to ammonia and isocyanic acid and continues the same process.

A diesel engine equipped with a diesel particulate filter and SCR system (with aqueous urea) meets ultra low emission vehicles (ULEV) emission standards (0.2 g/mi NOx, 0.04 g/mi PM) (Lambert et al., 2001). NOx emission decreased more than 80% in a diesel engine with an SCR system under a transient and steady state condition test (Conway et al., 2005). However, N_2O, which is a greenhouse gas, increased with an SCR system at low exhaust gas temperature (about 200°C) due to thermal decomposition of ammonium nitrate, but N_2O emission can be controlled by maintaining NO/NO_2 ratio 1 (Bartley and Sharp, 2012). A schematic diagram of SCR is shown in Figure 4.8.

4.16.3 Lean NOx Trap

A lean NOx trap (LNT) is an aftertreatment device used to reduce NOx emissions in a lean burn engine. NOx is stored in the LNT during the lean operation of an engine. When the air-fuel ratio becomes rich, the stored NOx is catalytically reduced by the reductants (CO, H_2, and HC) (Kim et al., 2003). A schematic diagram of LNT is shown in Figure 4.9.

4.16.4 Water Injection

The addition of water to the engine could decrease NOx emission significantly. Water is either injected into the intake manifold during a suction stroke or directly into the cylinder during combustion. NOx emission deceases with water mainly due to a reduction in combustion temperature as heat sink (due to high specific heat) and an increasing dilution effect. Besides NOx reduction, a water injection strategy provides increasing volumetric efficiency by creating a cooling effect and acting as a working fluid (mostly supercritical stage). Demineralized water is used for this purpose to avoid a corrosion problem in the engine components. Water can be emulsified with diesel, which is called a water-diesel emulsion or water-fuel emulsion. As water can't mix with diesel, it needs surfactant to

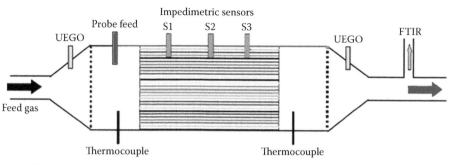

UEGO: universal exhaust gas oxygen sensor
FTIR: Fourier transform infrared spectroscopy analysis

FIGURE 4.9
Schematic diagram of a lean NOx trap.

make the emulsion. NO emission in a diesel engine with the emulsified fuel (diesel + water) at the rated load decreased from 752 to 463 ppm with a water-to-diesel ratio of 0.5:1 and it is reported that the optimum water-to-diesel ratio is from 0.4 to 0.5:1 (by mass) (Subramanian and Ramesh, 2001). Both methods (water injection and water emulsion) can reduce NOx emission at the source level, but the water emulsion method is more effective in terms of the quantum of NOx emission reduction because NO emission in a diesel engine is 398 and 477 ppm at 60% load with emulsion and injection, respectively (Subramanian, 2011). Smoke emission would decrease with a water-in-oil emulsion due to better mixing because water-impregnated diesel undergoes an unstable superheated state. The boiling point of water is less than diesel, and therefore the specific volume of water expands, resulting in an explosion of water particles called microexplosion, and hence, the surrounding diesel gets fine fragment and better mixing with air. In the case of water injection, smoke emission will increase. Water is also used for increasing the hydrogen energy share in a dual-fuel diesel engine (Chintala and Subramanian, 2016). A schematic diagram of water injection is shown in Figure 4.10.

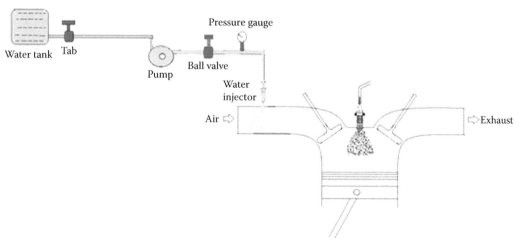

FIGURE 4.10
Schematic diagram of water injection.

4.16.5 Optimization of Injection Timing

NOx decreases with retarding injection timing as peak pressure and temperature are lower as they are shifted after TDC, which results in lower NOx formation. However, the opposite occurs as it increases with advanced injection timing. Similarly, spark timing retarding from maximum brake torque (MBT) timing, NOx emission is generally lower but is higher with advanced spark timing.

4.17 PM Emission Reduction Techniques

Controlling PM emission is a major challenge to researchers. While PM in modern engines is less than 0.1% of the fuel mass due to the advancement of technologies such as CRDI, PM on volume basis (number of count and size) is high. If PM size is less than 2.5 μm, the human respiratory system may be unable to filter the particles and they may deposit on the lungs. The World Health Organization defines diesel PM as having a carcinogenic nature and may lead to the growth of cancer cells in biological systems, including including that of humans. Therefore, PM in internal combustion engines needs to be reduced in order to meet the current stringent emission norms for sustaining the environment and ecological systems. PM emission forms in engines due to the following:

- Poor fuel quality, such as higher level of sulfur, higher density, or viscosity
- Poor mixture formation due to poor fuel spray, aerodynamic characteristics, and heterogeneous mixture
- Improper combustion phasing (lower premixed charge combustion)
- Lower in-cylinder temperature

PM emission can be reduced using the following strategies:

- Improving fuel quality
- Improving aerodynamic characteristics (swirl, tumble, squish, optimization of combustion chamber geometry)
- Improving fuel spray characteristics
- Optimization of design, operating, and combustion parameters
- Aftertreatment devices

4.17.1 PM Emission Reduction by Better Fuel Quality

Physicochemical properties such as density, viscosity, distillation characteristics, octane number/cetane number, and sulfur influence the mixing rate of the injected fuel with the surrounding air in the combustion chamber. For example, fuel with higher density and viscosity and less vaporization would reduce fuel spray cone angle and vaporization, and also increase Sauter mean diameter, resulting in poor mixing and hence high PM emission. Sulfur in fuel is a main source of PM emission. Lower octane number fueled engine has to operate at a lower compression ratio, resulting in a lower in-cylinder temperature and hence more nanoparticulates. Similarly, the distillation property of a fuel affects the vaporization process. If the spray cone angle is lower due to higher density and viscosity of fuel, the inner core of the fuel is a source of soot emission. Hence, fuel quality plays a vital role in PM emission, since the emission can be controlled by updating the physicochemical properties of the fuel.

4.17.2 PM Emission Reduction by Improving Aerodynamic Characteristics

Important aerodynamic characteristics such as swirl (organized rotation about the cylinder axis), tumble (organized rotation perpendicular to the cylinder axis), and squish (radial inward motion toward a piston's bowl at the end of a compression stroke) need to be improved by optimizing the geometry of an engine's intake system (improved design of swirl port/plate, manifold, and valve) and piston geometry (improved design of a piston's bowl geometry to enhance the turbulence field). The fuel spray orientation/direction and air fluid flow direction need to be synchronized for better mixing of the injected fuel with air.

4.17.3 PM Emission Reduction by Better Fuel Spray Characteristics

Important spray characteristics are breakup length, penetration distance (liquid penetration and vapor penetration), Sauter mean diameter (SMD), spray cone angle, air entrainment, and vaporization. Increasing fuel injection pressure in the range of (1500 to 2200 bar and even higher) will reduce SMD drastically and hence allow better mixing of the injected fuel with the surrounding air. Higher fuel injection pressure will increase the penetration distance and air entrainment. Fuel at the high-injection pressure rate is likely in the vapor phase in order to mix properly. The optimization of the nozzle hole diameter, number of holes, fuel rate shaping, and so forth, will also help to reduce the PM level significantly. Research work continues further in enhancement of diesel fuel injection pressure of more than 2200 bar to achieve the desired injected fuel droplet diameter, which should be almost equal to fuel gaseous molecule diameter such as diameter of oxygen molecule or gaseous molecule.

4.17.4 PM Emission Reduction by Optimizing Combustion Phase

The increase in a premixed charge may be difficult in a diesel engine due to combustion with knock. However, the premixed charge can easily be increased using partly gaseous fuel for reducing PM emission. PM emission at the source level can be significantly reduced using HCCI and RCCI modes. Low reactivity fuel such as ethanol and compressed natural gas can be used for enhancing a premixed charge in a diesel-fueled compression-ignition engine in RCCI mode and PM emission will be reduced drastically as well. PM emission in internal combustion engines can also be reduced effectively with engine operation in HCCI mode. An engine can be operated in HCCI mode by a combination of EGR, air heating, fuel rate shape (by varying injection rate, providing pulse injection), and optimizing the engine's design and operating parameters.

4.17.5 PM Emission Reduction by Design, Operating Parameters, and Combustion Phasing

Design parameters, such as compression ratio, variable valve timing, and combustion chamber geometry, and operating parameters, such as air-fuel ratio, speed, load, EGR rate, and oxygen enrichment (using membrane system), need to be optimized for reducing PM emission. For example, increasing the compression ratio in an SI engine would lead to high in-cylinder temperature, resulting in better oxidation of particles. The oxygen enrichment using a membrane system will also significantly reduce PM emission in diesel-fueled compression-ignition engines. If the speed of the CI engine increases, cycle time and mixture preparation time will be reduced, which may lead to undesirably high PM emission. Optimization of combustion characteristics, such as those for keeping a higher heat release rate and cumulative heat release, occurrence of the start of combustion of nearby TDC, 50% of heat release of nearby TDC, and lower combustion duration, are important for engine operation with less PM emission. Because a diesel engine's PM emission is lower with a lean fuel-air mixture, lean burn is an effective technology to reduce the emission. Therefore, the important parameters for PM reduction include increased premixed charge, high oxygen level, high mixture preparation time, high in-cylinder temperature, lean burn, and combustion phase optimization.

4.17.6 PM Emission Reduction by Aftertreatment Devices

PM in an exhaust manifold is separated using a physical filtration device called a particulate trap or diesel particulate filter (DPF). The trap is made of material such as ceramic monolith, woven silica fiber, ceramic foam, and wire mesh. The structure of the trap is generally either monolith or honeycomb with the inner surfaces coated by catalysts. Trapped carbon particles have to be regenerated by burning or removing them; otherwise, the trap's conversion efficiency will be reduced and back pressure of exhaust gas will increase.

Summary

A spark-ignition engine operates with an Otto cycle whereas a compression-ignition engine operates with a constant pressure cycle/dual cycle. The thermal efficiency of the engine increases with an increasing compression ratio and specific heat ratio. However, the compression ratio of spark-ignition engines is limited due to knock. In contrast, the compression ratio of compression-ignition engines are not thermodynamically limited but are practically limited based on mechanical/thermal stress on the materials of the engine components. The thermal efficiency of engines with the same design and operating conditions is the highest with an Otto cycle than with a constant pressure cycle or dual cycle. The efficiency increases with an increasing compression ratio, α (pressure ratio of heat addition at constant volume) and specific heat ratio; however, it decreases with an increasing cutoff ratio. In general, the compression ratio is higher with spark-ignition engines than with compression-ignition engines and hence the thermal efficiency of CI engines is higher. Spark-ignition engines provide a higher specific power output than compression-ignition engines because SI engines can operate with an equivalence ratio of up to one.

- A four-stroke engine is widely used in automotive and other applications as it gives better thermal efficiency and less emissions compared to two-stroke engines. Usage of two-stroke engines for automotive transportation has been almost eliminated due to poor thermal efficiency and high levels of emissions.

- The products from these engines with complete combustion (100% combustion efficiency) are CO_2 and H_2O vapor. However, practical engines emit emissions such as CO, HC, NOx, N_2O, PM, CH_4, and H_2.

- Even though the operating cost and maintenance cost of an internal combustion engine are the highest, it is still a preferable choice because the initial cost of the system per kilowatt is still the lowest compared to other heat engines.

- Twin driving forces for the development of internal combustion engines are thermal efficiency improvement and emissions reduction. The deciding factor for choosing the type of ignition engine is mainly based on fuel cost.

- NOx and PM emissions are higher in diesel-fueled compression-ignition engines. The modern diesel engine emits relatively less emissions along with improved thermal efficiency compared to the engines used in earlier decades. These engines are equipped with advanced technologies such as CRDI, variable valve timing, variable injection timing, variable fuel rate shaping (pulse injection), EGR, SCR/LNT, DOC, and PM trap.

- Advanced technologies such as the RCCI engine and HCCI are in the research and development (R&D) stage for improving performance of an engine with less emissions. Engines with these technologies can meet the desired emission and fuel economy norms.

- Diesel and gasoline are primarily used as fuels in internal combustion engines/vehicles. However, there are alternative-fueled vehicles. Biofueled vehicles get more attention due to their better performance and emission reduction compared to conventionally fueled vehicles.

- An important note is that the effectiveness of an engine technology depends mainly on fuel quality. For example, a technology for improving the thermal efficiency of an engine depends on the octane number of fuel. Therefore, both fuel quality and engine technology are the main elements for achieving better performance and emission reduction in internal combustion engines. In this context, since biofuels have better fuel quality than petroleum fuels, as seen in the previous chapters, the design and operating parameters of engines can be better modified for further performance improvement and emission reduction.

Solved Numerical Problems

1. A spark-ignition engine operates with an Otto cycle. The compression ratio of the engine is 10:1. Calculate air standard thermal efficiency.

Input Data

Compression ratio (r) = 10:1

Specific heat ratio for air (γ) = 1.4

Thermal efficiency (η_t) = ?

Solution

$$\eta_t = 1 - 1/r^{\gamma-1}$$
$$= \left(1 - 1/(10)^{(1.4-1)}\right) \times 100 = 60\%$$

2. A single cylinder spark-ignition engine has a 100-mm bore and 100-mm stroke that runs with 20 NM at 3000 rpm. Calculate brake power and brake mean effective pressure (BMEP).

Input Data

Bore of engine (d) = 100 mm

Stroke of engine (L) = 100 mm

Torque (T) = 20 NM

Speed (N) = 3000 rpm

Number of cylinders (K) = 1

Brake power (BP) = ?

$BMEP$ = ?

Solution

$$BP = \frac{2\pi NT}{60 \times 1000}$$

$$= 2 \times 3.14 \times 3000 \times 20/(60 \times 1000)$$

$$= 6.28 \text{ kW}$$

$$BMEP = \frac{BP \times 2 \times 60}{Vs \times N \times K \times 100}(bar)$$

Swept volume (Vs) $= 3.14 \times (0.1^2/4) \times 0.1 = 0.000785 \, \text{m}^3$

$BMEP = 6.28 \times 2 \times 60/(0.000785 \times 3000 \times 1 \times 100) = 3.2$ bar

3. A four-cylinder compression-ignition engine runs at 5000 rpm with 4-bar BMEP and the engine's swept volume is 2 liters. Calculate brake power output of the engine.

Input Data

Number of cylinders = 4

$BMEP$ = 4 bar

Speed = 5000 rpm

Swept volume = 2 liters = 0.0025 m³

Brake power output = ?

Brake power = $BEMP \times Vs \times N/2 \times K$

$$= 400000/1000 \times 0.002 \times 5000/(2 \times 60) \times 4$$

$$= 133.3 \text{ kW}$$

4. A diesel engine coupled with an alternator produces electrical power. The pole of the alternator is 4 and the frequency of electrical power is 60 Hz. Find out the speed of the alternator.

Input Data

Number of poles $(P) = 4$

Frequency $(f) = 60$ Hz

Speed of the engine $(N) =$?

Solution

$$N = 102 \times f / P$$

$$= 120 \times 60/4$$

$$= 1800 \text{ rpm}$$

5. A timber industry company needs electrical power with the following power demand with respect to time. Choose the proper heat engines.

   2 MW for 1 hour (4 pm to 5 pm)

   4 MW for 5 hours (8 am to 1 pm)

   8 MW for 3 hours (1 pm to 2 pm)

   12 MW for 2 hours (2 pm to 4 pm)

   The power grid will supply a few kilowatts required for lighting and other off-line work at night.

   The maximum power demand of the company is 12 MW and the minimum demand of power is 2 MW. If a heat engine delivers a rated power output of 12 MW, the load of the engine is 16%, 33%, 67%, 100%. The engine has to be operated with part load most of the time. The thermal efficiency of the engine is much less at part load than at the rated load. A preferable choice may be three internal combustion engines that will deliver an electrical power output of 4 MW. This is a better choice than a gas turbine because with a gas turbine, the power requirement is discrete and there is a fluctuating load. A steam turbine plant is not economically feasible for this type of low power demand.

   One engine with 4 MW with 100% load will supply electrical power for 5 hours

   Two engines with 8 MW with 100% load will supply electrical power for 3 hours

   Three engines with 12 MW with 100% load will supply electrical power for 2 hours

   One engine with 50% load will supply electrical power supply for 1 hour

Sustainable Approach: The timber industry company would have plentiful resources of wood waste. The waste disposal management needs energy input and other logistic. A better choice is that producer gas can be generated using wood waste and gas can be utilized as the fuel in a spark-ignition engine for power generation. The company would get affordable electrical power generation with reduced problems of waste disposal.

References

Bartley, G. and Sharp, C. (2012), Brief investigation of SCR high temperature N_2O production, *SAE International Journal of Engines*, 5(2),683–687, doi:10.4271/2012-01-1082.

Chintala, V. and Subramanian, K. A. (2016), Experimental investigation of hydrogen energy share improvement in a compression ignition engine using water injection and compression ratio reduction, *Energy Conversion and Management*, 108,106–119.

Conway, R., Chatterjee, S., Beavan, A., Lavenius, M. et al. (2005), Combined SCR and DPF Technology for Heavy Duty Diesel Retrofit, SAE Technical Paper 2005-01-1862, doi:10.4271/2005-01-1862.

Kalsi, S. S. and Subramanian, K. A. (2016), Experimental investigations of effects of EGR on performance and emissions characteristics of CNG fueled reactivity controlled compression ignition (RCCI) engine, *Energy Conversion and Management*, 130,91–105.

Kim, Y., Sun, J., Kolmanovsky, I. and Koncsol, J. (2003), A phenomenological control oriented lean NOx trap model, SAE Technical Paper 2003-01-1164, doi:10.4271/2003-01-1164.

Kuihua, H. A. N. and Chunmei, L. U. (2007), Kinetic model and simulation of promoted selective non-catalytic reduction by sodium carbonate, *Chinese Journal of Chemical Engineering*, 15(4),512–519.

Lambert, C., Vanderslice, J., Hammerle, R. and Belaire, R. (2001), Application of urea SCR to light-duty diesel vehicles, 2001-09-24, SAE 2001-01-3623.

Lin, Z., Takeda, K., Yoshida, Y., Iijima, A. et al. (2016), Influence of EGR on knocking in an HCCI engine using an optically accessible engine, SAE No. 2016-32-0012.

Robertson, D., Chadwell, C., Alger, T., Zuehl, J. et al. (2017), Dedicated EGR vehicle demonstration, SAE 2017-01-0648.

Røjel, H., Jensen, A., Glarborg, P. and Dam-Johansen, K. (2000), Mixing effects in the selective non-catalytic reduction of NO, *Industrial and Engineering Chemistry Research*, 39(9),3221–3232.

Subramanian, K. A. (2011), A comparison of water–diesel emulsion and timed injection of water into the intake manifold of a diesel engine for simultaneous control of NO and smoke emissions, *Energy Conversion and Management*, 52(2),849–857.

Subramanian, K. A and Ramesh, A. (2001), A study on the use of water-diesel emulsions in a DI diesel engine, SAE Technical Paper 2001-28-0005, doi:10.4271/2001-28-0005.

5

Basic Processes of Internal Combustion Engines

5.1 Spark-Ignition Engines

The basic processes of spark-ignition engines include mixture formation, ignition, combustion, and combustion products formation. A layout of these processes and how the relevant areas of science deal with them is shown in Figure 5.1. A spark-ignition engine converts chemical energy into heat and then mechanical energy by combustion. The mixture formation is an important process that deals with mixing of inducted air with injected/inducted fuel during a suction stroke. The air is inducted due to the difference in pressure between the in-cylinder and the atmosphere as the piston moves downward, resulting in creation of vacuum pressure inside the cylinder. The desired air-fuel ratio and its quantity have to be maintained according to the load change of the engine.

The next important process is ignition, which initiates the chemical reaction between the air and fuel. Activation energy is needed to break down molecules of air-fuel that is trapped between the electrodes in a spark plug. Activation energy is defined as the minimum amount of energy required for a species in order to undergo a specified reaction. The required ignition energy, which is more than activation energy, is provided by the spark plug. The flame kernel grows with the input of ignition energy provided by the spark plug. Then, the developed flame kernel initiates the combustion process, in which the chemical energy of fuel is converted into heat energy. The flame kernel can be defined as group of molecules of an air-fuel mixture in the vicinity of the spark plug, which has attained sufficient energy to ignite itself without any further aid of energy from the spark plug. The formation of the flame kernel occurs during the ignition delay period of the engine.

The heat conversion rate depends mainly on flame speed. Flame speed is mainly a function of the reactant pressure and temperature, equivalence ratio, spark intensity, charge density, and residual gas fraction. Ignition is generally an endothermic process as few air-fuel molecules trapped inside a spark plug gap get external energy provided by the spark plug.

In contrast, combustion is an exothermic process in which heat energy is released. Flame velocity influences the performance of spark-ignition engines. If the flame velocity is higher, the heat release rate would be higher, resulting in enhancement of the degree of constant volume combustion. Note that thermal efficiency increases with an increase in the degree of constant volume combustion. An increase in flame velocity increases the in-cylinder temperature and the quality of heat, resulting in better conversion from heat to work.

Products that result from the combustion in spark-ignition engines are carbon dioxide and water vapor if the combustion is complete or combustion efficiency is 100%. Nitrogen will leave as it is because it does not contribute to combustion. However, the combustion

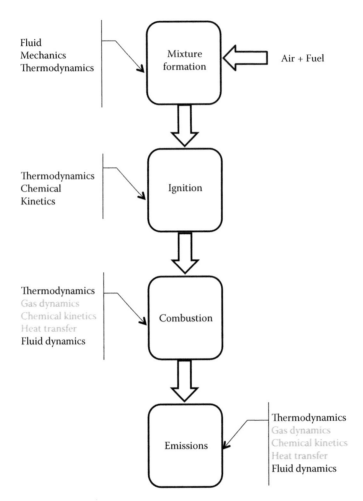

FIGURE 5.1
Spark-ignition engine processes.

efficiency of actual spark-ignition engines is less than 100% for various reasons, including poor air-fuel mixing, insufficient oxygen, low reactant pressure and temperature, low flame velocity, and flame quench. These engines emit emissions such as CO, HC, NOx, and nano-PM.

The important basic processes of spark-ignition engines and the related mathematical expressions for the relevant parameters in the processes are explained below.

5.1.1 Fuel Induction/Injection

High octane number fuels are used in spark-ignition engines. The fuels at atmospheric condition are generally highly volatile substances or in gaseous state. For example, gasoline fuel is in liquid state while it is stored in a fuel tank but its phase would change from liquid to vapor while it is inducted or injected to an engine's intake manifold, where the fuel is mixed with the inducting air. Similarly, liquefied natural gas (LNG), LPG, and liquefied hydrogen (LH) are in a liquid phase in their fuel tank. When these fuels come into contact with the inducted air, their phase changes to vapor or gas.

The fuel is inducted to an engine's cylinder in four ways:

1. Carburetor
2. Manifold injection
3. Port injection
4. Direct injection

5.1.1.1 Carburetor

A carburetor is a simple mixing device of air and fuel in an engine's intake manifold. This device does not need any energy input for its function and it operates due to the pressure difference between the in-cylinder and fuel. This device consists of a venturi, throttle valve, nozzle jet, and fuel container, and is connected in the engine's intake manifold. Ambient air that is inducted from the atmosphere flows from upstream to downstream the carburetor. Simultaneously, the fuel is fed to the venturi of the carburetor, and hence, both air and fuel gets mixed and then the charge enters into the engine's cylinder. A typical carburetor is shown in Figure 5.2.

The in-cylinder pressure is in the range from 0.4 bar to 0.9 bar during a suction stroke whereas normal temperature and pressure (NTP) of ambient air is 1.01325 bar and 20°C. However, ambient temperature and pressure varies with a change in altitude with

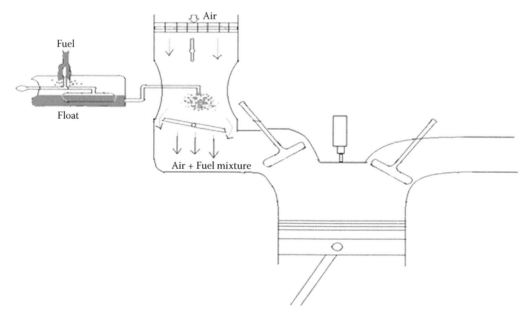

FIGURE 5.2
Schematic diagram of a carburetor.

reference to sea level and geographical zone. The mass flow rate of air can be calculated using multiplication of volume flow rate of air and its density, as shown in

$$m_a = \frac{P_a \times V_a}{R \times T_a} \tag{5.1}$$

where
 m_a = Mass flow rate of air (kg/s)
 V_a = Volume flow rate of air (m³/s)
 P_a = Ambient air pressure (N/m²)
 T_a = Ambient air temperature (K)
 R = Universal gas constant (8314 J/kg K)

The density of ambient air decreases with increase in its temperature, and vice versa. The ambient air pressure decreases with an increase in altitude, and vice versa. At a hill station, ambient pressure is relatively lower compared to a reference level of sea datum. On the other hand, ambient air temperature is also lower at a hill station. The density of ambient air at a hill station is lower because its pressure is lower, but in contrast, the density of air should increase as its temperature is lowered. The net effect of air density change depends on the dominant effect of temperature and pressure. In a hill station, the density of air would generally be lower than that of reference sea datum (surface over earth).

The mass flow rate of fuel depends on its density and volume flow rate, as shown in

$$m_f = V_f \times \rho_f \tag{5.2}$$

where
 m_f = Mass flow rate of fuel (kg/s)
 V_f = Volume flow rate of fuel (m³/s)
 ρ = Density of fuel (kg/m³)

The amount of fuel can be increased by the percentage of the throttle opening relative to the full throttle opening. In the case of a carburetor mode, the required quantity of fuel will flow with respect to the throttle opening position. The density of fuel can be increased by increasing its pressure using a fuel pump. The volume flow rate of fuel can be increased or decreased by adjusting jet size in a carburetor. If the density of fuel varies, the mass fuel rate will change.

5.1.1.2 Manifold Injection

Fuel is injected into an engine's intake manifold using an electronic injector. The intake manifold is a long pipe and its downstream side (bend shape) is attached to the engine's cylinder head, whereas its upstream side is attached with an air filter that is exposed to the atmosphere. The layout of a manifold injection is shown in Figure 5.3. The injector, which is a solenoid-operated device, is mounted on the intake manifold. The injector orientation (its location and angle) plays an important role in better mixing of air with fuels (Chintala et al., 2013). If the injector is located too close to the intake valve, mixing time is reduced. Durability issues with the injector can also occur due to the mounting point in the manifold having a high temperature, where if the injector is too far away from the intake valve,

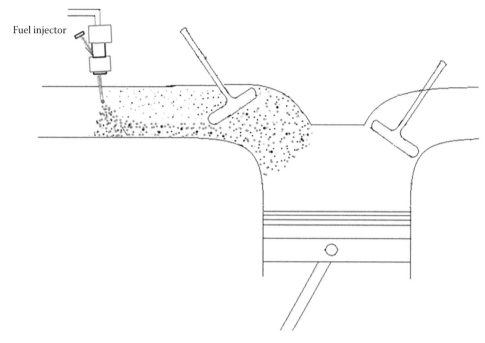

FIGURE 5.3
Schematic diagram of a manifold injection.

the charge will not enter completed into the engine's cylinder. The injector orientation needs to be optimized for better mixing of a charge. This method is preferable for gaseous fuel injection because there are no wall-wetting problems.

5.1.1.3 Port Injection

The injector is located nearby valve in the engine's intake valve. The injected fuel will directly enter into the in-cylinder through the intake valve. This method is preferable for liquid fuel or poor vaporization fuel; otherwise, these fuels may have the problem of wall-wetting if they are injected inside the manifold. A diagram of the port injection is shown in Figure 5.4.

5.1.1.4 Direct Injection

The injector is mounted on the cylinder head near the spark plug, as shown in Figure 5.5. The fuel is injected during a compression stroke or the closing of all valves. The main advantages of this direct injection method is significant improvement in volumetric efficiency and specific power output due to more air that is inducted by the engine during a suction stroke. A carburetor or manifold/port injection system supplies fuel and mixes with the inducting air but some amount of air is replaced by the fuel, resulting in less volumetric efficiency. This problem can be addressed by the direct injection method. However, this method has disadvantages because it requires a specialized injector for injecting fuel at high pressure. In addition, mixer preparation time would be reduced and the charge would be a type of stratification (less premixed/homogenous charge) compared to other injection methods.

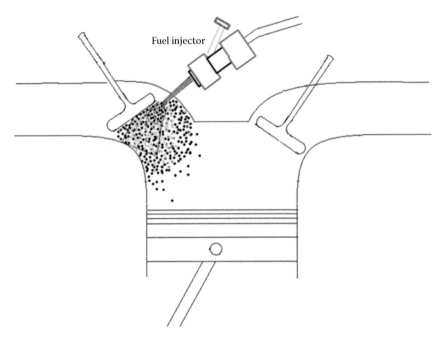

FIGURE 5.4
Schematic diagram of a port injection.

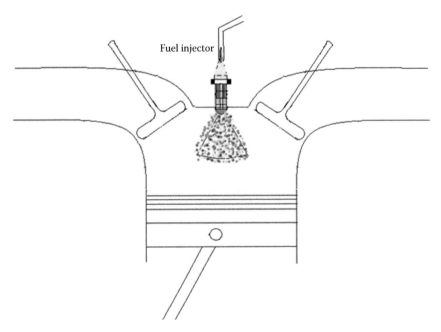

FIGURE 5.5
Schematic diagram of a direct injection.

5.1.2 Volumetric Air Efficiency

Volumetric air efficiency is defined as the ratio of the actual mass flow rate of air to the theoretical air flow rate. The theoretical volumetric efficiency in a naturally aspirated spark-ignition engine is 100%. However, actual volumetric efficiency is less due to air flow restrictions from mechanical devices such as a carburetor venture, intake valve (head and stem), and thermal parameters such as residual gas temperature and flow characteristics, backflow of exhaust gas into the intake manifold, a boundary layer's shear stress, and the effect of geometry (intake manifold and combustion chamber) on localized flow velocity. Volumetric air efficiency can be calculated using

$$\text{Volumetric efficiency } \eta_v = \frac{2\dot{m}_a}{\rho_{ai} \times V_s \times N} \tag{5.3}$$

where $\dot{m}_a$ is the mass flow rate of air, ρ_{ai} is the density of incoming air, V_s is the swept volume of the engine, and N is the speed of the engine.

The mass flow rate of air is a function of the density of air and the volume flow rate of air. The density of air varies with respect to atmospheric pressure, atmospheric temperature, and boost pressure by a supercharger or turbocharger and residual gas temperature. The volume flow rate of air depends on the swept volume and geometry of intake and valve systems. The mass flow rate of air is less at high altitude (hills) or high atmospheric temperature (desert) compared to standard conditions.

The volumetric efficiency of spark-ignition engines is less due to the throttling system compared to compression-ignition engines. Volumetric efficiency increases with an increase in speed but it decreases beyond a certain speed due to the mechanical limitation of the intake valve opening and closing at high engine speed. Volumetric efficiency can effectively be increased using a supercharger or turbocharger.

5.1.3 Charge Motion

5.1.3.1 Swirl

Swirl is an organized rotation of charge about a cylinder axis (Figure 5.6). Swirl enhances mixing of inducted/injected fuel with the surrounding air until the end of a compression stroke. The swirl ratio is the ratio of charge swirl to engine speed:

$$R_s = \frac{\omega_s}{2 \times \pi \times N} \tag{5.4}$$

where ω_s is the angular velocity, such as a solid-body rotating flow, which has equal angular momentum to the actual swirl flow.

5.1.3.2 Tumble

Tumble is an organized rotation of charge perpendicular to the cylinder axis (Figure 5.7). Tumble is more preferable than swirl in spark-ignition engines as the flame propagation will be faster with a tumble flow of charge because the unburned charge can better be

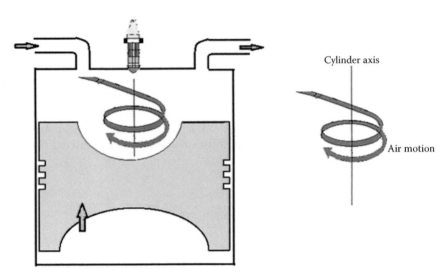

FIGURE 5.6
Schematic diagram of a swirl flow.

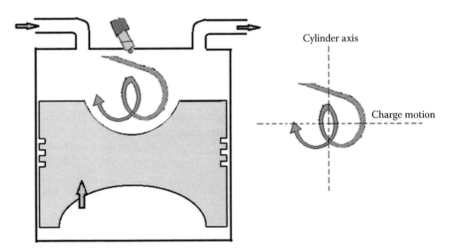

FIGURE 5.7
Schematic diagram of tumble flow.

transported from the bottom to the top of the cylinder during ignition and combustion so that the charge obtains a faster combustion. The tumble ratio can be defined as the ratio of tumble charge to engine speed, as shown in

$$R_t = \frac{\omega_t}{2\pi N} \tag{5.5}$$

where ω_t is the angular velocity of a solid-body rotating flow, which is equal to the angular momentum of the actual tumble motion of air.

5.1.4 Ignition

Ignition is an important chemical process in a spark-ignition engine. The chemical reaction between air and fuel is initiated by ignition.

5.1.4.1 Minimum Ignition Energy

Initiating ignition needs at least a minimum ignition energy (*MIE*), which can be described using

$$Q_{MIE} = Q_r - Q_G + Q_L \tag{5.6}$$

where
Q_{MIE} = Minimum ignition energy (kJ)
Q_r = Energy required for the reactant in a spark plug electrode domain to initiate chemical reaction (ignition) (kJ)
Q_G = Heat generation (kJ) (heat may be generated by the reactants during the ignition period)
Q_L = Heat loss (kJ) (heat lost to an electrode by conduction and surrounding reactants by convection

If heat generation and heat loss are assumed to be zero, Equation 5.6 can be written as Equation 5.7

$$Q_{MIE} = Q_r \tag{5.7}$$

where

$$Q_R = m_r \times C \times (T_f - T_r) = \rho \times C \times V \times (T_f - T_r) \tag{5.8}$$

The mass of the reactant is only considered for ignition in the region of space between spark plug electrodes. The volume of space between electrodes can be assumed as spherical. Equation 5.8 can be written in the form

$$Q_R = MIE = \rho \times C \times \frac{\pi \times d_{opt}^3}{6} \times (T_f - T_r) \tag{5.9}$$

where
ρ = Density of reactant (kg/m³)
C = Specific heat (kJ/kg K)
V = Volume of reactant (air + fuel) in a spark plug's electrode zone (m³)
d_{opt} = Optimum diameter of a sphere in the spark plug electrode zone (m); diameter will change with respect to electrode distance change
T_f = Flame kernel temperature (K)
T_r = Initial temperature of reactant (air + fuel) (K)

Note that the *MIE* requirement for spark-ignition engines vary with respect to electrode distance, as explained next.

MIE is lowest when the electrode distance is at an optimum. *MIE* is higher when the electrode distance is smaller or larger compared to the optimum distance, because when the electrode gap is less than optimum, heat transfer loss is too high, resulting in more ignition energy requirement. In contrast, when the gap is higher, the heat requirement is also high because a greater quantity of reactant needs to be heated, which will result in more ignition energy requirement.

5.1.4.2 Breakdown Energy from a Spark Plug

The life of a spark is categorized into three phases: breakdown phase, arc phase, and glow phase. The first two phases are very short. Breakdown voltage V_{bd} (kV) between electrodes is calculated using (Sjeric et al., 2015)

$$V_{bd} = 4.13 + 13.6(P/T) + 324(P/T)xd \qquad (5.10)$$

where
  V_{bd} = Breakdown voltage (kV)
  P = Instantaneous in-cylinder pressure (bar)
  T = Instantaneous in-cylinder temperature (K)
  d = Spark plug electrode gap (mm)

Using this breakdown voltage, breakdown energy (E_{bd}) can be calculated using

$$E_{bd} = V_{bd}^2 / C_{bd}^2 \ xd \qquad (5.11)$$

where
  E_{bd} = Breakdown energy (J)
  C_{bd} = Breakdown constant (V/(J.mm)$^{0.5}$)

The activation energy of molecules can be calculated using Equations 5.12 and 5.13 (Nautiyal et al., 2014).
  To calculate the reaction rate constant, the following formula has been adopted:

$$k = ln(UB_o) - ln(UB_t)/t \qquad (5.12)$$

where
  UB_o = Initial unburned mass of charge
  UB_t = Unburned mass after time t
  t = Time taken to complete one degree of crank angle (milliseconds)

By taking a logarithm of Arrhenius' equation ($k = Ae^{-E_a/(RT)}$), we get:

$$ln \ k = ln \ A - E_a/RT \qquad (5.13)$$

5.1.4.3 Diameter of a Flame Kernel

The diameter of a flame kernel can be calculated using Equations 5.14 to 5.16 (Salvi et al., 2015):

$$A_i = 2 \times \left[\frac{\Upsilon - 1}{\Upsilon} \times \frac{E_{bd}}{\pi P_o l_g \left(1 - \dfrac{T_o}{T_i} \right)} \right] \tag{5.14}$$

$$d_i = \frac{4}{\pi} \times \sqrt{A_i} \tag{5.15}$$

where

A_i = Area of flame kernel (mm²)
d_i = Diameter of flame kernel (mm)
Υ = Isentropic expansion factor
E_{bd} = Breakdown energy (kJ)
l_g = Distance between two electrodes (mm)
P_o = Instantaneous in-cylinder pressure (bar)
T_o = Instantaneous in-cylinder temperature (K)
T_{bd} = Breakdown temperature (840K)
T_i = Temperature of plasma channel (K)

$$T_i = \left[\frac{1}{\Upsilon} \left(\frac{T_{bd}}{T_o} - 1 \right) + 1 \right] \times T_o \tag{5.16}$$

5.1.5 Combustion

5.1.5.1 Instantaneous Cylinder Volume

The geometry of an engine's cylinder, piston, connecting rod, and crankshaft is shown in Figure 5.8. In the figure, V_d is the swept volume, V_c is the clearance volume, s is the distance between the piston pin and crank center, l is the connecting rod length, a is the crank radius (= stroke (L)/2), B is the bore diameter, and L is the stroke length.

The instantaneous cylinder volume at any crank position θ is given by

$$V(\theta) = V_c + \frac{\pi D^2}{4} (l + a - s) \tag{5.17}$$

where s

$$s = a \cos\theta + (l^2 - a^2 \sin^2\theta)^{1/2} \tag{5.18}$$

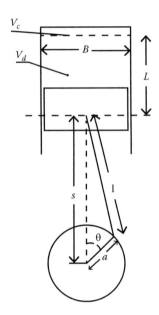

FIGURE 5.8
Geometry of an engine's piston-cylinder.

$$V(\theta) = V_c + \frac{\pi D^2}{4} \left[l + a - a\cos\theta - (l^2 - a^2 \sin^2\theta)^{\frac{1}{2}} \right] \tag{5.19}$$

$$V(\theta) = V_c \left\{ 1 + \frac{1}{2}(r_c - 1) \left[\frac{l}{a} + 1 - \cos\theta - \left(\frac{l^2}{a^2} - \sin^2\theta \right)^{\frac{1}{2}} \right] \right\} \tag{5.20}$$

where $r_c = (V_d + V_c)/V_c$ is the compression ratio of the engine.

5.1.5.2 Mean Piston Speed

The mean piston speed is calculated by

$$\bar{S} = 2NL \tag{5.21}$$

5.1.5.3 Instantaneous Piston Speed

The instantaneous piston speed is calculated by

$$\frac{S(\theta)}{\bar{S}} = \frac{\pi}{2} \sin\theta \left[1 + \frac{\cos\theta}{\left(\left(\frac{1}{a} \right)^2 - \sin^2\theta \right)^{1/2}} \right] \tag{5.22}$$

where l is the connecting rod length, a is the crank radius, and L is the stroke length.

5.1.5.4 In-Cylinder Temperature

The in-cylinder temperature is calculated by

$$T(\theta) = \frac{p(\theta)V(\theta)}{m(\theta)R} \tag{5.23}$$

where p is the in-cylinder pressure, V is the instantaneous cylinder volume, m is the instantaneous mass within the cylinder, and R is the gas constant.

5.1.5.5 Rate of Pressure Rise

Rate of pressure rise can be related indirectly to combustion with knock, smooth running of an engine, and failure and life of an engine. If rate of pressure rise is higher than the desired level (generally <6 bar pressure/degree crank angle), the above stated problem in an engine will occur.

$$\frac{dp}{d\theta} = p_2(\theta) - p_1(\theta) \; if \; one \; degree \; crank \; angle \; interval \tag{5.24}$$

5.1.5.6 Heat Release Rate

Heat release rate is an important combustion characteristic of an internal combustion engines. It can be calculated using the first law of thermodynamics with input of in-cylinder pressure, in-cylinder temperature, specific heat ratio, and instantaneous volume of an engine cylinder with respect to crank angle. Heat release rate indicates how much quantity of chemical energy is converted into heat per degree crank angle.

The important combustion characteristics of a spark-ignition engine such as flame development angle, rapid burning angle, and overall burning angle can be interpreted using the heat release rate data. The main combustion characteristics of a compression-ignition engine such as ignition delay, premixed combustion phase, diffusion combustion phase, and after-burning phase can be found out using the heat release rate data. In addition, other cumulative characteristics including heat release and combustion duration can also be calculated.

The total heat release during combustion is given by

$$\delta Q = dU + \delta W \tag{5.25}$$

$$\delta Q = mc_v dT + pdV \tag{5.26}$$

Then inserting $dT = \dfrac{pdV + Vdp}{mR}$ into Equation 5.26:

$$\delta Q = mc_v \left(\frac{pdV + Vdp}{mR} \right) + pdV \tag{5.27}$$

$$\delta Q = pdV\left(1 + c_v\left(\frac{1}{R}\right)\right) + \frac{c_v}{R}vdp \tag{5.28}$$

Then inserting $\dfrac{c_v}{R} = \dfrac{1}{\gamma - 1}$ into Equation 5.28:

$$\frac{dQ}{d\theta} = \frac{\gamma}{\gamma - 1}p\frac{dV}{d\theta} + \frac{1}{\gamma - 1}V\frac{dp}{d\theta} \quad \text{(without considering heat loss)} \tag{5.29}$$

$$\frac{dQ}{d\theta} = \frac{\gamma}{\gamma - 1}p\frac{dV}{d\theta} + \frac{1}{\gamma - 1}V\frac{dp}{d\theta} + Q_{loss} \quad \text{(considering heat loss)} \tag{5.30}$$

where γ is the ratio of specific heat and Q_{loss} is heat loss to the cylinder wall.

5.1.5.7 Mass Fraction Burned

The mass fraction burned can be calculated using the Wiebe function (Heywood, 1988) and is given by

$$x_b(\theta) = 1 - exp\left[-a\left(\frac{\theta - \theta_0}{\Delta\theta}\right)^{m+1}\right] \tag{5.31}$$

where θ_0 is the start of combustion, $\Delta\theta$ is the combustion duration, and a and m are adjustable parameters.

5.1.5.8 Correlation for Convective Heat Transfer Coefficient

Heat is released inside a cylinder during combustion and then the in-cylinder temperature increases. The heat from the combustion chamber through the conduction mode is transferred to the cylinder wall and then it is transferred to the atmosphere through radiation and convection mode. The quantification of the wall heat transfer through conduction, convection, and radiation mode is complex, because overall heat transfer coefficient needs to be determined using the Nusselt number with input of the Reynolds and Prandtl numbers. Woschni's correlation is widely used to calculate convective heat transfer coefficient, as given in (Woschni, 1967)

$$h_c = 0.82 \times B^{-0.2}p^{0.8}T^{-0.53}w^{0.8} \tag{5.32}$$

where B is the bore and w is the average gas velocity in the cylinder.

5.1.5.9 Turbulent Flame Speed

The turbulent flame is characterized by the root mean square velocity fluctuations, turbulence intensity, and various lengths of scales of the turbulent flow ahead of the flame.

Turbulent flame speed depends on flow conditions as well as mixture properties, and is given by (Herweg and Maly, 1992):

$$\frac{S_t}{S_l} = I_0 + \sqrt{I_0} \times \left[\frac{(U^2 + u^2)^{\frac{1}{2}}}{(U^2 + u^2)^{\frac{1}{2}} + S_l} \right]^{\frac{1}{2}} \times \left[1 - \exp\left(-\frac{r}{l} \right) \right]^{\frac{1}{2}}$$

$$\times \left[1 - \exp\left(-\frac{(U^2 + u^2)^{\frac{1}{2}} + S_l}{l} \times t \right) \right]^{\frac{1}{2}} \times \left(\frac{u}{S_l} \right)^{\frac{5}{6}} \tag{5.33}$$

where

$$Io = 1 - \left(\frac{\delta_l}{15l} \right)^{0.5} \times \left(\frac{u}{S_l} \right)^{1.5} - 2 \times \frac{\delta_l}{r} \times \frac{\rho_u}{\rho_b}$$

$$l = 0.7 \times V_u^{\frac{1}{3}}$$

$$u = 0.65 \times S_p = 0.65 \times 2LN$$

$$\delta_l = \left(\frac{k}{\rho C_p} \right)_u \frac{1}{S_l}$$

5.1.5.10 Laminar Flame Speed

The laminar speed can be defined as the velocity at which a flame propagates into a premixed unburned mixture. Flame is the result of a self-sustaining chemical reaction occurring within a region of space called the flame front, where unburned mixture is heated and converted into products. Laminar flame speed depends on only the thermal and chemical properties of the mixture.

The laminar flame speed for gasoline is given by (Heywood, 1988)

$$S_L = S_{L,O}(T_u/T_o)^\alpha (P/P_o)^\beta \tag{5.34}$$

where

$$\alpha = 2.18 - 0.8(\varphi - 1) \tag{5.35}$$

$$\beta = -0.16 + 0.22(\varphi - 1) \tag{5.36}$$

$$S_{L,O} = B_m + B_\varphi(\varphi - \varphi_m)^2 \tag{5.37}$$

where T_u is the unburned mixture temperature, T_o is the ambient temperature, P_o is the ambient pressure, S_t is the turbulent flame velocity (cm/s), S_l is the laminar flame velocity (cm/s), ρ_u is the unburned charge density (g/cc), and ρ_b is burned charge density (g/cc).

5.1.5.11 Combustion Efficiency

Hydrocarbon fuel gets thermal pyrolysis during combustion and intermediate products form, including carbon monoxide, smaller hydrocarbon fragments, and radicals. The chemical reaction continues until the main product of carbon dioxide and water vapor form, because carbon in fuel is converted to carbon dioxide, whereas hydrogen in fuel is converted to water vapor. The combustion efficiency of an engine fueled with hydrocarbon fuel would be 100% when the engine emits only CO_2 and H_2O as products. If intermediate products, such as CO, unburned and partially burned hydrocarbon, and soot, present in the engine's exhaust system, the combustion efficiency of the engine is less than 100%. The combustion efficiency of an engine is calculated using

$$\eta_{comb} = \left(1 - \frac{(m_a + m_f)}{m_f \times CV_f} \sum_i x_i CV_i\right) \times 100 \tag{5.38}$$

where x_i and CV_i are the mass fraction and calorific value of species i in exhaust, respectively.

5.1.5.12 Combustion Duration

The combustion duration in terms of crank angle can be defined as the duration between the start and end of combustion. The end of combustion can be predicted from the crank angle corresponding to 90% of the cumulative heat release. The combustion duration can be described as given in

$$\text{Combustion duration} = \int_{SOC}^{90\% \text{ heat release}} d\theta \tag{5.39}$$

5.1.6 Performance

5.1.6.1 Indicated Power

Indicated power is defined as the actual power developed inside the cylinder of an engine and can be calculated using

$$IP = \frac{IMEP \times V_s \times N \times k}{n \times 60} \tag{5.40}$$

where
   IMEP = Indicated mean effective pressure = work done/swept volume
   V_s = Swept volume
   N = Speed (rpm)
   n = Number of strokes
   k = Number of cylinders

5.1.6.2 Brake Power

Brake power is defined as the useful power available at the crankshaft of the engine and can be calculated using

$$BP = IP - FP \tag{5.41}$$

$$BP = \frac{2\pi NT}{60} \tag{5.42}$$

$$BP = \frac{BMEP \times V_s \times N \times k}{n \times 60} \tag{5.43}$$

where *FP* is the frictional power, *T* is the torque, and *BMEP* is the brake mean effective pressure.

5.1.6.3 Thermal Efficiency

The thermal efficiency of an engine is the ratio of power output to chemical energy input of the engine and can be calculated using

$$\eta_{th} = \frac{BP}{\dot{m}_f CV} \tag{5.44}$$

where $\dot{m}_f$ is mass flow rate of fuel and *CV* is calorific value of the fuel.

5.1.7 Emissions

5.1.7.1 CO Emission

Carbon monoxide is a derivative of incomplete combustion and is a partially burned fuel. CO emission depends mainly on the air-fuel equivalence ratio, which is usually of low significance during steady state diesel engine operation. If the mixture (air-fuel) does not have enough oxygen present during combustion, fuel will not burn completely. As combustion takes place in a lower oxygen surrounding, there is insufficient oxygen present to fully oxidize the carbon atoms in CO_2. The lower oxygen surrounding occurs due to a richer air-fuel mixture. There are several engine operating conditions, such as during cold-start operation, warm-up, and power enrichment, when this occurs normally.

5.1.7.2 Hydrocarbon Emission

Hydrocarbon emission forms due to incomplete combustion, which is dependent on many parameters including poor mixing, rich fuel-air mixture, ultramixture (beyond flammability limit), lower in-cylinder temperature, postoxidation of fuel trapped in crevice volume, and flame quenching. Total hydrocarbon (THC) is the summation of nonmethane hydrocarbon (NMHC) and methane and are given in a later section (Section 9.9.2 in Chapter 9).

5.1.7.3 NOx Emission

NOx is a mixture of mainly NO and NO_2 gases. In most high-temperature combustion processes, the majority (95%) of NOx formation is in the form of NO. NOx formation is primarily a function of in-cylinder pressure and temperature, residence time of combustion products, premixed combustion, availability of oxygen, ignition delay period, and the operational parameters of the engine. NOx forms through: thermal NO, fuel-bound NO, and prompt NO. Thermal NO is a postflame phenomena that can be analyzed using the extended Zeldovich mechanism, as given in Equations 5.45 through 5.47. Thus, NO formed in postflame gases would be much higher than NO formed in the flame.

Fuel-bound NO is formed by the reaction of fuel-bound nitrogen and oxygen at temperatures more than 850°C and is mostly dependent on the availability of oxygen and the combustion method. Nitrogen oxides formed as a result of the faster conversion of atmospheric nitrogen and oxygen in the flame front during combustion is called prompt NO.

$$N_2 + O \leftrightarrow NO + N \tag{5.45}$$

$$N + O_2 \leftrightarrow NO + O \tag{5.46}$$

$$N + OH \leftrightarrow NO + H \tag{5.47}$$

Thus the rate of formulation of NO is

$$\frac{d[NO]}{dt} = k_1[O][N_2] + k_2[N][O_2] + k_3[N][OH] - k_4[NO][N] - k_5[NO][O] - k_6[NO][H] \tag{5.48}$$

where, k_1, k_2, and k_3 are the forward rate constants and k_4, k_5, and k_6 are the reverse rate constants.

Conversion of emissions on a volume-to-mass basis:

$$\dot{m}_{e,i} = \frac{x_i \times MW_i \times \dot{m}_{ex} \times CF \times 3600}{MW_{ex} \times BP} \tag{5.49}$$

where

$\dot{m}_{e,i}$ (kg/kW-hr) = Specific mass emission of species i (I = CO, HC, NOx, CO_2, CH_4, N_2O, etc.)

x_i = Volume fraction of emission i (i = %/100 (or) ppm $\times 10^{-6}$)

MW_i = Molecular weight of emission i

$\dot{m}_{ex}$ = Mass flow rate of exhaust gas

MW_{ex} = Molecular weight of exhaust gas

CF = Correction factor that depends on the relative humidity, temperature, and pressure of ambient air

The correction factor (CF) is given by (EPA, electronic code of federal regulations)

$$CF = K_w \times K_H \, (K_H = 1 \text{ for CO and HC emissions}) \tag{5.50}$$

where

$$K_H = \frac{1}{1 - 0.0182(H - 10.71)} \tag{5.51}$$

$$\text{Humidity ratio} \, (HR) = \frac{622 \times P_v}{P_{amb} - P_v} \, (\text{g of vapor per kg of dry air}) \tag{5.52}$$

$$K_w = (1 - FFH \times FAR_{dry}) - K_{wl} \tag{5.53}$$

where

$$FFH = ALF \times 0.1448 \times \left(\frac{1}{1 + FAR_{dry}} \right) \tag{5.54}$$

$$\text{Hydrogen mass percentage of the fuel} \, (ALF) = \frac{1.008\alpha}{12.01 + 1.008\alpha} \tag{5.55}$$

where α is the H/C mole ratio of the fuel

$$\text{Dry fuel-air ratio} \, (FAR_{dry}) = \frac{m_f}{m_{a,wet} \times \left(1 - \frac{HR}{1000} \right)} \tag{5.56}$$

and

$$K_{wl} = 1.608 \left(\frac{HR}{1000 + 1.608\,H} \right) \qquad (5.57)$$

5.2 Compression-Ignition Engines

The basic processes of compression-ignition engines include injection, mixture formation, combustion, and emissions. A layout of these processes and how the relevant areas of science deal with them is shown in Figure 5.9.

5.2.1 Fuel Injection Characteristics

The mass flow rate of fuel injected through a nozzle is given in

$$\dot{m}_f = C_d A_n \sqrt{2 \rho_f \Delta p} \qquad (5.58)$$

where A_n is the minimum area of the nozzle, C_d is the discharge coefficient, ρ_f is fuel density, and Δp is the pressure drop across the nozzle.

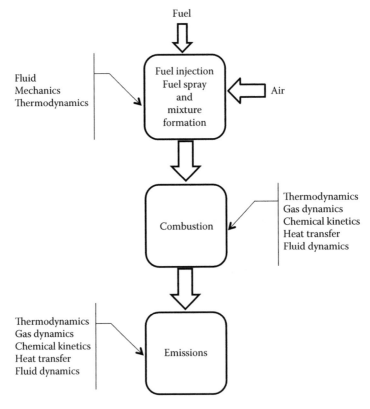

FIGURE 5.9
Processes of a compression-ignition engine.

5.2.2 Fuel Spray Characteristics

5.2.2.1 SMD

The mean diameter of a spray droplet is used to describe the quality of atomization. SMD (D_{sm}) is the diameter of a droplet that has the same surface-to-volume ratio as that of the total spray, as given in

$$D_{sm} = \left(\frac{\sum n_i \times x_i^3}{\sum n_i \times x_i^2} \right) \tag{5.59}$$

where x_i is the droplet diameter and n_i the is number of droplets found that are the same size.

The high viscosity and surface tension of biodiesel would increase the SMD, resulting in poor atomization (Lee et al., 2005):

$$D_{sm} = A(\Delta p)^{-0.135} (\rho_a)^{0.121} (V_f)^{0.131} \tag{5.60}$$

where A is a constant that depends on the type of nozzle and V_f is the amount of fuel delivered per cycle per cylinder in cubic millimeters per stroke.

5.2.2.2 Spray Breakup Length

The distance a jet moves from a nozzle outlet before breaking up in a spray is generally called the breakup length. The breakup length is reduced as the pressure difference across a nozzle is increased. A smaller breakup length indicates better air entrainment and lower ignition delay. The breakup length is influenced by the nozzle hole diameter, air density and turbulence, and the physical characteristics of the fuel. The correlation of breakup length is shown in (Dan, 1997)

$$L_b = 15.8 \times \left(\frac{\rho_f}{\rho_a} \right)^{0.5} \times D_n \tag{5.61}$$

where D_n is the nozzle diameter (m).

5.2.2.3 Spray Cone Angle

The injected fuel disintegrates after the breakup length due to surface tension and aerodynamic interaction with air and the fuel stream. A liquid fuel is injected at high injection pressure (generally in the range of 300–2200 bar) for better atomization of fuel particles. The sprayed particles are suspended in a combustion chamber of an engine from an injector tip to the surface of a piston bowl. Thus, this spray plum appears like a cone shape, and fuel particles diverge from the tip of the injector hole. Spray cone angle is an angle covered between the core of spray and a peripheral edge of spray. Spray cone angle is used for better air entrainment into the spray, resulting in better mixing of fuel with air. The increase in pressure difference across the nozzle results in an increase in spray cone angle.

Spray cone angle is also influenced by air density and the physical parameters of the fuel (Araneo et al., 1999).

$$\tan\frac{\theta}{2} = \frac{1}{A} \times 4\pi \times \left(\frac{\rho_f}{\rho_a}\right)^{0.5} \times \frac{3^{0.5}}{6} \tag{5.62}$$

where $A = 3 + \left[\dfrac{\left(\dfrac{L_n}{D_n}\right)}{3.6}\right]$ and L_n is the nozzle length.

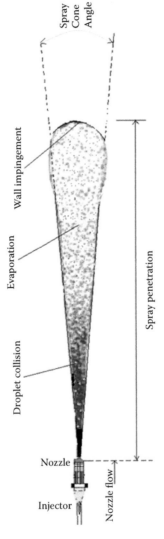

FIGURE 5.10
Schematic diagram of spray characteristics.

5.2.2.4 Spray Penetration

Spray penetration (Dent, 1971) is a measure of the depth of the spray and it indicates how much distance the spray traces before the fuel droplets are vaporized. A high spray penetration would be beneficial for more air entrainment into a fuel spray, but it may enhance the probability of wall impingement. However, poor fuel spray decreases the degree of homogeneity of an air-fuel mixture and it may lead to a fuel-rich region or a region that is too lean, which could lead to combustion with an increased degree of heterogeneous mixture or poor oxidation with an ultralean mixture in the combustion chamber. Poor spray penetration may lead to undesirable combustion, which could lead to poor engine performance and emission. The spray characteristics are shown in Figure 5.10.

$$S = 3.07 \left(\frac{\Delta P_{inj}}{\rho_g} \right)^{1/4} (Dt)^{1/2} \left(\frac{294}{T_g} \right)^{1/4} \tag{5.63}$$

where, ΔP_{inj} is the pressure drop across the injector, D is the nozzle diameter, t is the time, ρ_g is the gas density, and T_g is the gas temperature.

5.2.2.5 Air Entrainment

Air entrainment is the process of air drawn into the fuel spray so that homogeneity of the fuel-air mixture in the combustion chamber is maintained. Air entrainment plays a vital role in the mixing process of air and fuel. Fuel is injected at the end of compression in diesel engine so that compressed air enters tangentially into the injected fuel spray, which is called air entrainment. The air entrainment process is mainly dependent on the spray cone angle and spray penetration. Therefore, fuel spray characteristics are a critical part of the design of any diesel engine. The air entrainment of fuel spray can be found using (Rakopoulos et al., 2004)

$$m_a = \frac{\pi}{3} \times \left[\tan \frac{\theta}{2} \right]^2 \times S^3 \times \rho_a \tag{5.64}$$

where s is the penetration distance.

Injection and spray characteristics are mainly dependent on fuel properties (viscosity, density, distillation characteristic, surface tension, specific heat, latent heat of vaporization, bulk modulus, etc.) and injection system characteristics (fuel injection pressure, injection duration, start of injection, fuel rate shaping, diameter of injector hole, number of injector holes, number of injection pulses, etc.).

5.2.3 Mixing Process: Swirl and Squish

5.2.3.1 Swirl Ratio

Swirl is defined as the organized rotation of air about a cylinder axis. This is an important aerodynamic characteristic that influences the mixture formation process. Swirl ratio is related to swirl flow and engine speed, as given in

$$R_s = \frac{\omega_s}{2\pi N} \qquad (5.65)$$

where ω_s is the angular velocity of a solid-body rotating flow, which has equal angular momentum to the actual swirl flow.

5.2.3.2 Squish

Squish is defined as the radial inward motion of air into the bowl of a piston at almost the end of a compression stroke (a few degrees crank angle before TDC) (Figure 5.11). The squish velocity for different types of bowl profiles is given by Equations 5.66 and 5.67 (Heywood, 1988).

Bowl-in-piston chamber:

$$\frac{v_{sq}}{S_p} = \frac{D_B}{4z}\left[\left(\frac{B}{D_B}\right)^2 - 1\right]\frac{V_B}{A_c z + V_B} \qquad (5.66)$$

Simple wedge chamber:

$$\frac{v_{sq}}{S_p} = \frac{A_s}{b(Z+c)}\left(1 - \frac{Z+c}{C+Z}\right) \qquad (5.67)$$

where V_B is the volume of the piston bowl, A_c is the cross-sectional area of the cylinder, S_p is the instantaneous piston speed, z is distance between the piston crown top and the

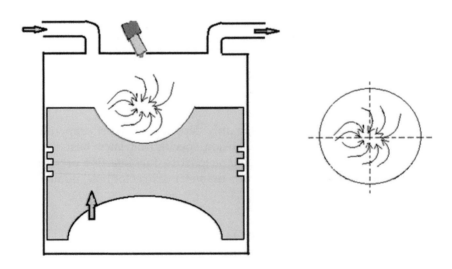

FIGURE 5.11
Schematic diagram of squish.

cylinder head ($z = c + Z$, where $Z = 1 + a - s$), A_s is the squish area, b is the width of the squish region, and C is $Z/(r_c - 1)$.

5.2.4 Ignition

Ignition is an important endothermic chemical process in a diesel engine. In diesel engines, fuel is injected into the engine cylinder during the end of the compression stroke. Combustion does not occur instantly with injected fuel, because a diesel engine needs a minimum amount of time for the mixture preparation process. Liquid fuel is usually injected at high pressure through small nozzles in the injector. It atomizes as small droplets and penetrates into the combustion chamber. The atomized fuel absorbs heat from the surrounding compressed air. Then, it vaporizes and mixes with the surrounding high-temperature and high-pressure air. The time lapse between the start of the injection and the start of combustion is known as ignition delay.

5.2.5 Combustion

5.2.5.1 Premixed Charge Preparation Rate

The premixed charge preparation rate can be calculated using

$$\dot{m}_{premixed,DI} = \int_{\theta_{SOI}}^{\theta_{SOC}} C_1 \left(\frac{\dot{m}_{diesel,DI}}{d\theta} \right)^{(1-C_2)} \left(\frac{\dot{m}_{diesel,unprepared}}{d\theta} \right)^{C_2} (p_\theta)^{C_3} \, d\theta, \text{ if } \theta_{SOC} > \theta_{SOI} \tag{5.68}$$

$$\dot{m}_{premixed,DI} = 0, \text{ if } \theta_{SOC} < \theta_{SOI} \tag{5.69}$$

The energy is supplied by the premixed and manifold injected charges:

$$E_{premixed,DI} = h_{fuel\,1} \times \dot{m}_{premixed,DI} \tag{5.70}$$

$$E_{premixed,MI} = h_{fuel\,2} \times \dot{m}_{fuel\,2,MI} \tag{5.71}$$

where C_1 (= 0.037), C_2 (= 0.36), and C_3 (= 0.54) are constants for controlling the charge preparation rate.

5.2.5.2 Percentage of Total Premixed Charge Energy

Premixed charge can relate indirectly the degree of homogeneous mixing of air-fuel, premixed combustion phase, rate of pressure rise, combustion with knock, and peak pressure. The percentage of total premixed charge energy can be calculated using Equation 5.72.

$$TPE = \frac{E_{premixed,DI} + E_{premixed,PI} + E_{premixed,MI}}{h_{fuel\,1}(\dot{m}_{premixed,DI} + \dot{m}_{fuel\,1,PI}) + (\dot{m}_{fuel\,2,MI} \times h_{fuel\,2})} \times 100 \qquad (5.72)$$

5.2.5.3 Degree of Homogeneity

The degree of homogeneity or heterogeneity can be calculated using

$$\text{Degree of homogeneity} = \frac{1}{n}\sum_{i}^{n}\varnothing_i \qquad (5.73)$$

$$\text{Degree of heterogeneity} = 1 - \text{Degree of homogeneity} \qquad (5.74)$$

where ϕ_i is the equivalence ratio of zone i.

5.2.5.4 Adiabatic Flame Temperature

Adiabatic flame temperature is defined as a system that can attain a maximum temperature for a given fuel-air mixture during combustion. The enthalpy of reactants and products increase with increasing temperature and vice versa. The adiabatic flame temperature can be determined by the condition that the enthalpy of the reactants is equal to that of products. The enthalpy used for this calculation is absolute enthalpy, which is determined by a summation of the enthalpy of the formation of elements/compounds and sensible enthalpy. For example, a stoichiometric equation for a hydrogen reaction with oxygen is given in Equation 5.75 and the adiabatic flame temperature of the hydrogen-air mixture can be calculated using Equations 5.76 and 5.77. The adiabatic flame temperature of any fuel-air mixture can be determined using this methodology.

$$H_2 + 0.5O \rightarrow H_2O \qquad (5.75)$$

$$\text{Enthalpy of Reactants}\,(H_R)\sum_{i}^{n}H_{R,i} = \sum_{i}^{n}H_{P,i} = \text{Enthalpy of Products}\,(H_P) \qquad (5.76)$$

$$H_{f_{H2}}^{O} + 0.5\,H_{f_{O2}}^{O} = H_{f_{H2O}}^{O} + C \times (T_{ad} - T^{O}) \qquad (5.77)$$

where
  H_f^o = Enthalpy of formation at standard conditions (Pressure : 1.01325 bar and Temperature (T^0) = 298 K)
  T_{ad} = Adiabatic flame temperature
  C = Specific heat $\left(\dfrac{kJ}{kg\text{-}k}\right)$

5.2.6 Emissions

5.2.6.1 Soot

Soot is a carbonaceous particle impregnated with tar material that consists of mostly amorphous carbon. A primary soot particle is constituted by thousands of crystallites. The density of a soot particle is about 2 gm/cm^3. Soot particles form PAHs when a flame is at a low-pressure condition. The formed soot can also mix with lubricating oil during combustion, causing the engine frictional power to increase due to contamination of the lubricating oil, as shown in Figure 5.12.

Incomplete combustion may result in soot formation in compression-ignition engines. Soot forms during combustion and subsequently oxidizes. The soot formation rate is always higher than the oxidation rate. Hence, the net soot oxidation rate is the difference between the soot formation rate and the soot oxidation rate. The net soot oxidation rate can be expressed by subtracting soot formation and its oxidation rate, as shown in

$$\frac{dm_s}{dt} = \frac{dm_{sf}}{dt} - \frac{dm_{so}}{dt} \tag{5.78}$$

Hiroyasu et al. (1983) proposed an empirical formula for calculating the soot formation and oxidation rates:

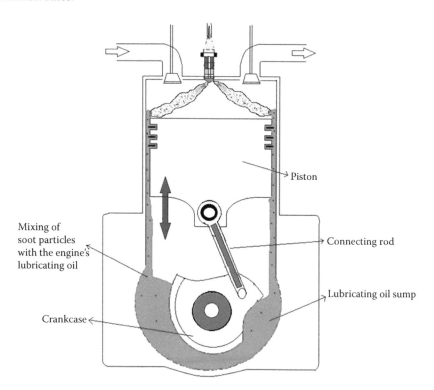

FIGURE 5.12
Schematic diagram of soot particles mixing with an engine's lubricating oil.

$$\frac{dm_{sf}}{dt} = A_f \times m_{fg} \times P^{0.5} \times \exp\left(\frac{-E_{sf}}{R \times T}\right) \tag{5.79}$$

$$\frac{dm_{so}}{dt} = A_c \times m_s \times \frac{P_{O2}}{P} \times P^{1.8} \times \exp\left(\frac{-E_{sc}}{R \times T}\right) \tag{5.80}$$

where
  A_f and A_c = Constants that are to be determined by matching the calculated amount of
      soot with experimentally measured soot in exhaust.
  m_{fg} = Mass of vaporized fuel (kg)
  P and T = In-cylinder pressure (MPa) and in-cylinder temperature (K)
  E_{sf} = 1.25 × 10⁴ kcal/kmol = The activation energy for soot formation

Stoot oxidation is predicted by assuming a second-order reaction between soot, ms, and oxygen.

  E_{sc} = 1.40 × 10⁴ kcal/kmol = The activation energy for soot oxidation
  m_s = Mass of soot in the package (kg)
  P_{O2} = Partial pressure of oxygen (MPa)

5.2.6.2 Smoke

Smoke is a carbonaceous particle suspended in an air column. Smoke forms due to a rich mixture and incomplete combustion. It is in the category of PM. Smoke is a visible indicator of carbon particles. If smoke is high, PM is high. If PM is high, smoke is not necessarily high as PM comprises organic and inorganic substance.

5.2.6.3 Particulate Matter

Particulate Matter (PM), which comprises organic and inorganic substances, forms in internal combustion engines due to incomplete combustion with rich mixture pockets in the combustion chamber. The particles are liquid, solid, or a combined form state (e.g., soot, smoke, dust, dirt). Certain heavier particles, such as soot, would not suspend in an air column, whereas smoke will. Soot and smoke contain only carbon or carbonaceous material. However, PM comprises organic and inorganic substances. PMs are classified based on their size, as given below:

  PM_{10}: Generally 105 µm (micrometers) or less and these particles are inhalable by
      human lungs
  $PM_{2.5}$: Inhalable particles generally 2.5 µm and smaller

Particles that are less than 10 micrometers can get deep into the lungs and even the bloodstream. Fine particles ($PM_{2.5}$) are the main cause of reduced visibility (haze) (EPA, Particulate Matter (PM) Pollution).

5.3 Analysis of the IC Engine Process Using Models and Simulation

An internal combustion engine process is influenced by one or more physical variables and the relationship between the physical variables can be presented in the form of mathematical equations. For example, the physical parameters for influencing the torque of an engine can be related in a mathematical form, as shown in Equation 5.81, and the torque can be calculated using an equation with input of *BMEP*, swept volume (Vs). The z, b, c, d, e, and f are constant. *BMEP* is a function of thermal efficiency (η_t, volumetric efficiency (η_v), fuel-air ratio (F/A), and air density, and the fuel's calorific value (CV) and their relationship is shown in Equation 5.82. Many physical parameters influence thermal efficiency (operating and design parameters), volumetric efficiency (cold-flow characteristics, effect of intake manifold and valve geometry, residual gas fraction and temperature, engine design and operating parameters, etc.), and air density (residual gas temperature, intake system geometry, etc.). Analysis of torque under wider operating and design parameters using experimental data is complex because experimental study of the effects of these parameters on torque is a time consuming process that is also expensive and complex. The developed mathematical equation called the model equation can be validated using only few experimental data and then the model equation can be used for analyzing the torque under wider operating conditions, resulting in a reduction in experimental time and minimizing expensive experimental tests.

$$T = \frac{BMEP \times V_s}{4 \times \pi} \tag{5.81}$$

$$BMEP = z \times \eta_t^b \times \eta_v^c \times F / A^d \times \rho_a^e \times CV^f \tag{5.82}$$

5.3.1 Zero-Dimensional Model

The model for analyzing engine processes can be classified into three categories, as detailed next.

5.3.1.1 Classification of the Zero-Dimensional Model

Zero dimension means that the effects of a fluid flow field on the domain are not considered. These models are mostly used for developing empirical models. The zero-dimensional model is mostly based on a thermodynamic equation such as the first law and second law of thermodynamics and chemical kinetics (reaction rate). The zero-dimensional model can further be classified into three categories:

1. *Single-Zone Model*: The instantaneous properties of working fluid is assumed to be the same in the entire zone. Examples of single-zone, zero-dimensional models are ignition delay, in-cylinder pressure, and heat release rate.

2. *Two-Zone Model*: There are two different distinct properties in two zones. For example, in a spark-ignition engine, a two-zone model (burned zone and unburned zone) is used for analyzing the working fluid using a thermodynamic equation. Another example for use of the two-zone model is analysis of flame kernel growth, in which one zone is considered for flame kernel growth in the vicinity of a spark plug and another zone is for unburned charge in the entire combustion chamber.

3. *Multi-Zone Model*: There are a number of zones to be considered for analysis of working fluid properties. For example, a three-zone model is used for analyzing the air entrainment process in which the surrounding air is considered for one zone, air entrained fuel is considered for a second zone, and the third zone is only for the fuel zone (inner core of spray).

5.3.1.2 Quasi-Dimensional Model/Phenomenological Model

The process of an internal combustion engine is analyzed in the entire zone with consideration of fluid flow effect in a specific tiny domain in the zone. For example, flame kernel growth between electrodes in spark plug can be analyzed using phenomenological model which is basically a thermodynamic model with considering the fluid flow effect only in the spark plug zone. Thus, the accuracy of the quasi-dimensional model/phenomenological model is higher than that of a zero-dimensional model.

5.3.1.3 Multidimensional Model

The effects of a fluid dynamic along with a thermodynamic model on an internal combustion engine process are analyzed using a multidimensional model approach. The mass, momentum, and energy conservation equations (Equations 5.83, 5.84, and 5.85) can be solved using an advanced computing system and a finite difference method (FDM)/finite volume method (FVM)/finite element method (FEM). The entire domain of the cylinder is discretized into a number of small zones or cells. The size of the cell is optimized based on the type of parameters (e.g., fuel droplet) to be predicted. The number of cells can be optimized using a grid independent study. If the number of cells are too high, the accuracy of the data is higher, but the time taken to converge the required solution is also higher. Depending on the required solution, computational time may take from a few hours to a few weeks.

The governing equation for mass conservation in the x-direction is given in

$$\frac{\partial \rho}{\partial t} + \frac{\partial \rho u}{\partial x} = 0 \tag{5.83}$$

The governing equation for momentum conservation in the x-direction is given in

$$\frac{\partial \rho u}{\partial t} + \frac{\partial uv}{\partial x} = -\frac{\partial P}{\partial x} + \frac{\partial \sigma_{ij}}{\partial x} \tag{5.84}$$

where ρ is density, u is velocity, P is pressure, and σ_{ij} is the stress tensor.

The governing equation for energy conservation in the x-direction is given in

$$\frac{\partial \rho e}{\partial t} + \frac{\partial u \rho e}{\partial x} = -P\frac{\partial u}{\partial x} + \sigma_{ij}\frac{\partial u}{\partial x} + \frac{\partial}{\partial x}\left(K\frac{\partial T}{\partial x} \right) \qquad (5.85)$$

where e is the specific internal energy, K is thermal conductivity, and T is temperature.

Summary

Engine performance, combustion, and emissions characteristics of internal combustion engines can be predicted using mathematical models and/or correlations available in the literature. These equations are mostly developed for petroleum-fueled engines. The basic equations may also be used for analyzing performance of biofueled engines but accuracy of data calculated using these model/correlation equations need to be validated using some experimental data. In addition, the constants of some equations may need to be tailored to work for biofueled engines. Basic simulation work can be performed using these equations for understanding the effects of biofuels on engine performance and emission formation.

The one-dimensional model is used for analyzing performance and combustion characteristics, such as volumetric efficiency, specific heat, ignition delay, and pressure. These parameters are analyzed using thermodynamics and the effects of fluid dynamics (aerodynamics) on these parameters are not generally taken into account. If these parameters need to be analyzed in depth—for example, flame kernel growth near a spark plug, air motion around an intake valve, or wall impingement on the surface area of a piston bowl—a phenomenological model is used for analysis of the parameters with inclusion of fluid dynamic effects in a particular (niche zone) domain area.

For example, the analysis of spray characteristics of biofuels in simulated/actual engine conditions is a complex process because of the effects of air motion (swirl and squish), heat transfer (from air to spray), and endothermic and exothermic reaction (thermodynamics and chemical kinetics) on the fuel spray. This analysis can be done using a two- or three-dimensional model with the help of dedicated computational processes that function mainly based on mass, momentum, and energy conservation. The required domain (plenum) is generally divided into the number of tiny cells to be analyzed using the FDM or FVM. The cell size is optimized based on accuracy and computational time.

A basic simulation is generally performed for assessment of the effect of the parameters on the entire engine's processes (performance, combustion, and emissions) under wider operating conditions (air-fuel ratio, engine speed, torque, spark timing, injection timing, etc.). The effects of design parameters, including intake manifold configuration, compression ratio, piston configuration, and valve timing on the engine processes can effectively be analyzed using a simulation. If the operating and design parameters are optimized using the simulation, some simulation data must be validated with the experimental data for confirming the accuracy of the simulation results. In this way, expensive and time-consuming experiments can be reduced or eliminated, resulting in development of technology or products in less time. A product that is created using simulation is called a virtual product.

Results from experimental tests conducted on internal combustion engines for measurement of their injection, spray, performance, combustion, and emissions can be analyzed and interpreted using the equations given in this chapter.

References

Araneo, L., Coghge, A., Brunello, G. and Cossali, G. E. (1999), Experimental investigation of gas density effect on diesel spray penetration and entrainment, International Congress & Exposition, Detroit, Michigan, SAE NO.1999-01-052:679-693.

Chintala, V. and Subramanian K. A. (2013), A CFD (computational fluid dynamics) study for optimization of gas injector orientation for performance improvement of a dual-fuel diesel engine, *Energy*, 57, 709–721.

Dan, T., Yamamota, T., Senda, J. and Fujimoto, H. (1997), Effect of nozzle configurations for characteristics of on reacting diesel fuel spray. International Congress & Exposition, Detroit, Michigan, SAE NO.970355:581-596.

Dent, J. C. (1971), Basis for the comparison of various experimental methods for studying spray penetration, SAE Paper 710571.

EPA, Electronic Code of Federal Regulations, Title 40, Part 89, Control of emissions from new and in-use non-road CI engines, http://www.ecfr.gov/cgi-bin/text-idx?SID=93528909feda1f23869b872807fd040f&mc=true&node=pt40.22.89&rgn=div5#se40.22.89_1112.

EPA, Particulate Matter (PM) Pollution, https://www.epa.gov/pm-pollution/particulate-matter-pm-basics#PM.

Herweg, R. and Maly, R. (1992), A fundamental model for flame kernel formation in S. I. engines, SAE Technical Paper 922243.

Heywood, J. B. (1988), *Internal Combustion Engine Fundamentals*, New York: McGraw-Hill.

Hiroyasu, H., Kadota, T. and Arai, M. (1983), Development and use of a spray combustion modelling to predict diesel engine efficiency and pollutant emissions, *Bulletin of the Japan Society of Mechanical Engineers*, 26(214), 569–575.

Nautiyal, P. and Subramanian, K. A. (2014), Kinetic and thermodynamic studies on biodiesel production from *Spirulina platensis* algae biomass using single stage extraction-transesterification process, *Fuel*, 135, 228–234.

Rakopoulos, C. D., Rakopoulos, D. C., Giakoumis, E. G. and Kyritsis D. C. (2004), Validation and sensitivity analysis of a two zone diesel engine model for combustion and emissions prediction, *Energy Conversion and Management*, 45(9–10), 1471–1495.

Salvi, B. L. and Subramanian, K. A. (2015), Experimental investigation and phenomenological model development of flame kernel growth rate in a gasoline fuelled spark ignition engine, *Applied Energy*, 139, 256–278.

Sjeric, M., Kozarac, D. and Taschl, R. (2015), Modelling of early flame kernel growth towards a better understanding of cyclic combustion variability in SI engines, *Energy Conversion and Management*, 103, 895–909.

Woschni, G. (1967), A universally applicable equation for the instantaneous heat transfer coefficient in the internal combustion engine, SAE Technical Paper 670931.

6

Utilization of Biofuels in Spark-Ignition Engines

6.1 Introduction

The main biofuels for spark-ignition engines are

Liquid biofuels:
- Methanol
- Ethanol
- Butanol

Gaseous biofuels:
- Biogas
- Hydrogen

6.2 Liquid Biofuels

Alcohol fuel is called octane booster because the octane number of gasoline fuel can be increased by blending it with methanol, ethanol, or butanol.

6.2.1 Advantages of Biofuel Use in Spark-Ignition Engines

The general advantages of alcohol fuels in spark-ignition engines are listed below.

1. The compression ratio of spark-ignition engines with alcohol fuels can be increased due to a higher octane number (RON of methanol: 112, ethanol: 111, and butanol: 113 (Rice et al., 1991)) than that of gasoline, resulting in higher thermal efficiency. As the antiknock index, which is defined by the average of the RON and MON, is higher with alcohol fuels, the probability of combustion with knock will be less, hence the engine will run smoothly and last longer. Engine's knock is an undesirable combustion phenomenon. If combustion proceeds with knock, in-cylinder pressure and temperature will increase rapidly and if the knock intensity is beyond threshold level, the knock may result in damaging the engine's components. If the flame velocity at a high knock intensity condition is more than sonic velocity, it is called detonation. In an SI engine, one of the reason for knocking is due to the end gas suppression by the flame. The unburned charge gets heated through convective

heat transfer by the flame. If the charge temperature is higher than the self-ignition temperature of the fuel, the unburned charge (end gas mixture) causes autoignition, which may result in knock. Knock with audible noise could damage the engine components, including valves, pistons, and its ring. An alcohol-fueled engine can operate without knock, relatively increasing the life of the engine.

2. The flame velocity of alcohol fuel is higher than gasoline fuel, which results in a better combustion rate, higher thermal efficiency, and less emissions. Flame velocity is directly proportional to mass burning rate, as shown in Equation 6.1. The in-cylinder pressure and temperature increase with an increase in the mass burned fraction, and hence the heat release rate is higher, whereas combustion duration is less. Flame velocity is a main combustion characteristic that influences the performance and emissions characteristics of spark-ignition engines. Note that reactant pressure and temperature increase with an increasing compression ratio of spark-ignition engines, resulting in higher flame velocity. As the octane number of alcohol fuels are generally higher than that of petroleum gasoline, a spark-ignition engine has the opportunity to increase its compression ratio for improving performance and reducing emissions. The degree of constant volume combustion increases due to higher flame speed with alcohol fuel that leads to a decrease in combustion duration, resulting in higher thermal efficiency. Therefore, alcohol-fueled engines could provide higher thermal efficiency.

$$\dot{m}_b = a_f \times \rho_u \times S_L \qquad (6.1)$$

where

$\dot{m}_b$ = Mass burning rate (kg/s)
a_f = Flame area (m^2)
ρ_u = Unburned charge density (kg/m^3)
S_L = Laminar burning velocity (m/s)

3. The latent heat of the vaporization of alcohol (methanol: 1100 kJ/kg, ethanol: 900 kJ/kg and butanol: 579 kJ/kg (Rice et al., 1991)) is higher than that of gasoline (400 kJ/kg). The volumetric efficiency of the engine would increase with alcohol fuel due to the intake's charge cooling because the temperature of air will decrease due to alcohol fuels having a higher latent heat of vaporization. The relationship between mass of air and temperature of air is shown in Equation 6.2. Latent heat means the fluid gets heat that is either being received or lost at a constant temperature. Therefore, when the inducted air-gasoline charge is mixing with the residual gas during a suction stroke, the temperature of the air initially increases due to heat transfer from hot residual gas (left over from the previous cycle) to the air. In the case of alcohol fuel, the charge temperature does not increase due to the fuel's higher latent heat of vaporization. Thus, since the mass flow rate of air increases with alcohol fuels, the volumetric efficiency of air in the engine will also increase (Equation 6.3) so that more fuel can be burned in order to get more power output.

$$\dot{m}_a = \dot{v}_a \times \rho_a = \dot{v}_a \times \frac{P_a}{R \times T_a} \qquad (6.2)$$

$$\eta_v = \frac{2\dot{m}_a}{\rho_{ai} V_s N} \qquad (6.3)$$

where

$\dot{m}_a$ = Mass flow rate of air (kg/s)
v_a = Volume flow rate of air (m³/s)
ρ_a = Density of air (kg/m³)
P_a = Pressure of air (kN/m²)
R = Universal gas constant (kJ/kg-K)
T_a = Temperature of air (K)
V_s = Swept volume of the engine (m³)
N = Speed of the engine (rpm)

4. Alcohol fuel is called oxygenated fuel because oxygen is embedded with hydrocarbon ($C_x H_y O_z$). The oxygen content in methanol, ethanol, and butanol is 49.9%, 34.7%, and 21.6% (by mass), respectively. Fuel containing oxygen would enhance the reaction rate of intermediate species (CO, partially burned hydrocarbon, etc.), which are effectively converted into the final products (CO_2 and H_2O). Therefore, the emissions of CO and HC will be less with alcohol fuels but it depends on the operating parameters and degree of mixing with air.

5. Reid vapor pressure is less with alcohol fuels, so volatile organic component (VOC) emission is less during filling at a gas station.

6. The carbon-to-hydrogen ratio (methanol: 3, ethanol: 4 and butanol: 4.8, and gasoline: 5.6 to 7.4 on a mass basis) is lower with alcohol than gasoline fuels so carbon-based emissions with alcohol fuels will be lower.

6.2.2 Disadvantages of Alcohol Fuel Use in Spark-Ignition Engines

1. The power output of engines with alcohol fuels is relatively less compared to that of gasoline-fueled engines due to the lower calorific value of alcohol fuels (methanol: 20.1 MJ/kg, ethanol: 27 MJ/kg, butanol: 33 MJ/kg, gasoline: 43.5 MJ/kg). The volumetric heat content of alcohol fuel (methanol: 15.9 MJ/liter, ethanol: 21.3 MJ/liter, butanol: 26 MJ/liter) is also less than that of gasoline fuel (gasoline: 32 MJ/liter) (Rice et al., 1991). If an engine needs to maintain the same power output (assuming the same thermal efficiency with alcohol and gasoline fuels), it needs to induct a relatively higher quantity of alcohol fuel (2.16 kg of methanol, 1.61 kg of ethanol, 1.31 kg of butanol compared to 1 kg of gasoline). However, a conventional engine may not be able to induct the fuel, resulting in a power drop. The power drop can be related to the mass of fuel and calorific value (same thermal efficiency for both fuels), as given in

$$\Delta BP = \eta(m_g \times CV_g - m_{acl} \times CV_{acl}) \qquad (6.4)$$

where m_g and CV_g are mass flow rate and calorific value of gasoline, respectively, and m_{acl} and CV_{acl} are mass flow rate and calorific value of alcohol fuel, respectively, and η is the thermal efficiency of the engine.

2. A spark-ignition engine fueled with alcohol fuel will have the problem of cold-start ability because this fuel has a higher latent heat of vaporization. Vaporization of this fuel will also depend on ambient temperature because problem of cold-start ability of the engine/vehicle is more in winter than summer.

3. NOx emission in spark-ignition engines could be less with alcohol fuel than with gasoline fuel. However, the oxygen content in the fuel tends to have more NOx formation but a higher latent heat of vaporization of alcohol tends to reduce emission formation. The latent heat of vaporization may be the dominant parameter, resulting in less NOx emission. However, other parameters may also contribute, including advanced or retarded spark timing from maximum brake torque (MBT) timing, occurrence of peak temperature near/before/after TDC during combustion, degree of homogenous air-fuel mixture in combustion chamber and localized temperature distribution, and air-fuel ratio.

The general effects of alcohol fuels on performance and emissions characteristics of spark-ignition engines have been discussed above. The distinct features of individual alcohol fuels are discussed next.

6.3 Utilization of Methanol in Spark-Ignition Engines

Performance of Spark-Ignition Engines Fueled with Methanol: The thermal efficiency of a spark-ignition engine fueled with methanol at 2500 rpm increases from 22% with a compression ratio of 6:1 (torque: 9 NM) to 30.5% with a compression ratio of 10:1 (torque: 10.52 NM) due to higher flame speed of methanol, less compression work due to higher heat of vaporization, and the presence of oxygen in methanol (Çelik et al., 2011). The efficiency of the engine with the same compression ratio (6:1) also increased from 19% with base gasoline to 22% with methanol. The brake power with methanol at a compression ratio of 6:1 dropped from 2.4 kW with base gasoline to 2.3 kW with methanol due to lower calorific value of methanol, but the power can be increased to 2.8 kW with methanol if the compression ratio was increased from 6:1 to 10:1. The compression ratio of the engine for methanol can be increased to 10:1 due to methanol having a higher octane number; however, the compression ratio of the engine is limited to 6:1 for gasoline due to knock because gasoline has a lower octane number. Hence, optimization of the compression ratio is necessary for a spark-ignition engine with methanol in order to get the desired power output and better thermal efficiency.

Emissions Characteristics of a Spark-Ignition Engine Fueled with Methanol: At the same compression ratio (6:1), CO and NOx emissions of the engine are less with methanol than that of gasoline, whereas HC emission is higher with methanol than with gasoline. CO emission is less due to better combustion by the presence of oxygen in the fuel. The reasons for low NOx and increased HC emissions are due to lower in-cylinder temperature. CO_2 emission is lower with methanol due to lower C/H ratio of 3 (by mass) than gasoline (5.6 to 7.4).

6.4 Utilization of Ethanol in Spark-Ignition Engines

Performance of Spark-Ignition Engines with Ethanol: The indicated thermal efficiency of a spark-ignition engine at a compression ratio of 11.5:1 and engine speed of 1500 rpm increased from 27% with base gasoline to 30.8% with ethanol due to the fast burning of ethanol, less compression work due to higher heat of vaporization, and presence of oxygen in ethanol (Turner et al., 2011). The power output at a compression ratio of 6:1 dropped from 1.9 kW with base gasoline to 1.82 kW with ethanol due to the lower calorific value of ethanol (Celik, 2008). The compression ratio of the engine for ethanol can be increased due to a higher octane number of ethanol and thus the power drop can be reduced with ethanol. Ethanol has a higher latent heat of vaporization, which leads to difficultly in starting the engine (cold-start) when the ambient temperature is low under cold conditions (Mani Sarathy, 2014). Alcohol fuels have a higher sensitivity (S) (S = RON – MON), and this high sensitivity is advantageous for engines that operate under LTC and direct fuel injection (Amer et al., 2012).

Emissions Characteristics of Spark-Ignition Engines Fueled with Ethanol: At the same compression ratio (6:1), CO and NOx emissions of an engine are less with ethanol than that of gasoline, whereas HC emission is higher with ethanol than with gasoline (Celik, 2008). CO emission is less due to better combustion from the presence of oxygen in the fuel. The reasons for lower NOx and higher HC emissions are due to lower in-cylinder temperature. CO_2 emission is also lower with ethanol due to lower C/H ratio of 4 (by mass) than gasoline (5.6 to 7.4) and improved combustion with ethanol.

6.5 Utilization of Butanol in Spark-Ignition Engines

Performance of Spark-Ignition Engines with Butanol: The indicated thermal efficiency of a single-cylinder CFR engine at a compression ratio of 10:1 and spark timing 10 crank angle before top dead center (^{0}bTDC) increased from 35.5% with base gasoline to 36.9% with butanol due to the fast burning rate of butanol, less compression work due to higher heat of vaporization, and the presence of oxygen in butanol (Szwaja et al., 2010). The MBT of an engine at a compression ratio of 9.6:1 and engine speed of 3000 rpm dropped from 55 Nm with base gasoline to 52 Nm with butanol due to the lower calorific value of ethanol (Gu et al., 2012). The compression ratio of an engine for ethanol can be increased due to a higher octane number of butanol, and thus, the power output can be increased with butanol.

Emissions Characteristics of Spark-Ignition Engines Fueled with Butanol: At the same compression ratio (9.6:1) and full load, NOx emission of an engine was less with butanol than with gasoline, whereas CO and HC emissions were higher with butanol than with gasoline (Gu et al., 2012). CO and HC emissions were higher due to reduced oxidation caused by lower in-cylinder temperatures. Low NOx emission with butanol was due to lower in-cylinder temperature and lower adiabatic flame temperature of butanol.

TABLE 6.1

Comparison of Emissions in Alcohol-Fueled SI Engines

Reference	Engine and Test Details	Fuel	CO	HC	NOx	% Reduction in CO	% Reduction in HC	% Reduction in NOx
Rice et al. (1991)	Four-cylinder SI engine, r = 8.5, Vs = 2.2 L, N = 2200 rpm, Load = full (115 joules) at WOT	M20	6.44 g/mile	9.67 g/mile	3.64 g/mile	88.83	12.88	11.43
		E20	16.7 g/mile	10.1 g/mile	3.81 g/mile	71.05	9.00	7.29
		B20	34.7 g/mile	10.9 g/mile	3.91 g/mile	39.86	1.80	4.86
		Gasoline	57.5 g/mile	11.1 g/mile	4.11 g/mile	–	–	–
Shenghua et al. (2007)	Three-cylinder SI engine, Vs = 0.8 L, N = 2500 rpm	M10	0.5%	870 ppm	2990 ppm	9.09	-1.16	0.33
		Gasoline	0.55%	860 ppm	3000 ppm	–	–	–
Al-Hasan (2003)	Four-cylinder SI engine, r = 9.0, Vs = 1452 cc, N = 3000 rpm	E 20	1.85%	140 ppm	–	46.37	26.31	–
		Gasoline	3.45%	190 ppm	–	–	–	–
Celik (2008)	Single-cylinder SI engine, r = 6.0, Vs = 250 cc, N = 2000 rpm, MBT timing and WOT	E25	2.65%	280 ppm	440 ppm	29.33	16	17.75
		E50	2.1%	245 ppm	390 ppm	44	26.8	27.1
		E75	1.3%	340 ppm	345 ppm	65	-1.49	35.5
		E100	0.75%	490 ppm	300 ppm	80	-46.2	43.9
		Gasoline	3.75%	335 ppm	535 ppm	–	–	–
Gu et al. (2012)	Three-cylinder SI engine, r = 9.4, Vs = 796 cc, N = 3000 rpm, Load = full (WOT), MBT timing	B 100	25 g/kWh	2.1 g/kWh	5.8 g/kWh	-38.8	-50	44.76
		Gasoline	18 g/kWh	1.4 g/kWh	10.5 g/kWh	–	–	–
Broustail et al. (2012)	Single-cylinder SI engine, r = 9.5, Vs = 499 cc, IMEP = 5 bar	E25	28.5 g/kWh	2500 ppm	8.6 g/kWh	3.38	16.6	1.14
		E50	27.5 g/kWh	2150 ppm	8.45 g/kWh	6.77	28.3	2.87
		E75	29 g/kWh	1650 ppm	8 g/kWh	1.69	45	8.04
		E100	25.4 g/kWh	1800 ppm	7.35 g/kWh	13.89	40	15.5
		B25	29.5 g/kWh	2600 ppm	8.95 g/kWh	0	13.33	-2.87
		B50	29.3 g/kWh	2250 ppm	8.85 g/kWh	0.67	25	-1.72
		B75	28 g/kWh	2150 ppm	8.7 g/kWh	5.08	28.3	0
		B100	31.9 g/kWh	2000 ppm	8.48 g/kWh	-8.13	33.3	2.52
		Gasoline	29.5 g/kWh	3000 ppm	8.7 g/kWh	–	–	–
Celik et al. (2011)	Single-cylinder SI engine, r = 6.0, Vs = 250 cc, N = 2000 rpm, MBT timing and WOT	M100	2.05%	480 ppm	290 ppm	25.4	-37.14	36.26
		Gasoline	2.75%	350 ppm	455 ppm	–	–	–
Turner et al. (2011)	Single-cylinder SI engine, r = 11.5, Vs = 565.6 cc, N = 1500 rpm, IMEP = 3.4 bar, ST = 24 °bTDC	E85	34 g/kWh	2.3 g/kWh	1.0 g/kWh	20.9	66.1	50
		E100	27 g/kWh	1.9 g/kWh	1.1 g/kWh	37.2	72	45
		Gasoline	43 g/kWh	6.8 g/kWh	2.0 g/kWh	–	–	–

6.5.1 Comparison of Performance and Emissions of Different Spark-Ignition Engines with Alcohol Fuels

A comparison of the emissions (CO, HC, and NOx) emitted from alcohol-fueled spark-ignition engines is given in Table 6.1. The emissions from alcohol- (methanol, ethanol, and butanol/with blends) fueled engines with different power outputs and design and operating parameters, were compiled and are given in Table 6.1. It can be observed and concluded from the table that CO, HC, and NOx emissions are mostly less with alcohol/alcohol-gasoline blends than with base gasoline fuel.

6.6 Utilization of Raw Biogas in Spark-Ignition Engines

Typical composition of raw biogas is approximately 60% methane and approximately 40% carbon dioxide with traces of H_2S, nitrogen, and hydrogen. The use of raw biogas as a fuel in spark-ignition engines would lead to the following problems.

Power Drop: As the calorific value of raw biogas (28 MJ/kg) is less than that of base gasoline, more biogas (approximately 1.5 times higher than gasoline (by mass)) needs to be inducted by the system in order to maintain the same power output. However, a conventional (unmodified) SI engine is not able to induct such a higher quantity of biogas, resulting in a power drop. Figure 6.1 shows the variation of brake power with respect to different CO_2 content in biogas and the equivalence ratio. The power drop increases with an increasing CO_2 percentage in biogas and is the highest with 37% CO_2 in biogas. The power output in all cases is the highest near the stoichiometric air-fuel ratio (equivalence ratio of 1) and is lower at both lean and rich mixtures.

Less Thermal Efficiency: Figure 6.2 shows the variation of brake thermal efficiency with respect to equivalence ratio and different CO_2 content in biogas. The brake

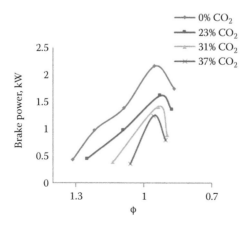

FIGURE 6.1
Variation of brake power with respect to the equivalence ratio. (From Bhatnagar, V. (2016), Study of performance and emissions characteristics of biogas fuelled spark-ignition engine, M.Tech. thesis, IIT Delhi.)

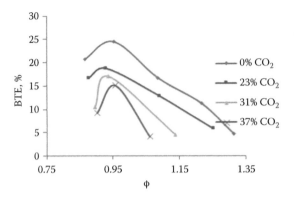

FIGURE 6.2
Variation of brake thermal efficiency with respect to equivalence ratio. (From Bhatnagar, V. (2016), Study of performance and emissions characteristics of biogas fuelled spark-ignition engine, M.Tech. thesis, IIT Delhi.)

thermal efficiency decreases with an increase in CO_2 in biogas mainly due to lower flame velocity and less adiabatic flame temperature. The flame velocity of raw biogas is lower than the pure substance of methane/gasoline fuel. The flame velocity of biogas with respect to equivalence ratio is given in Figure 6.3. It can be observed from the figure that the flame velocity increases with an increase in the equivalence ratio until stoichiometric ($\emptyset = 1$) and then it starts dropping.

The reduced flame velocity is due to CO_2 present in biogas. A flame with biogas cannot reach a cylinder wall due to its lower flame velocity. The unburned charge available in the unburned zone will not oxidize properly, resulting in poor combustion efficiency and thermal efficiency with a significant rise in CO and HC emissions. The flame velocity of raw biogas can be enhanced by increasing the compression ratio of the engine and

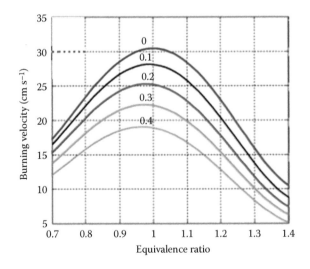

FIGURE 6.3
Laminar burning velocity with different molar CO_2 fractions (xCO_2) at T = 400 K and P = 5 bar. (From Hinton, N. and Stone, R. (2014), Laminar burning velocity measurements of methane and carbon dioxide mixtures (biogas) over wide ranging temperatures and pressures, *Fuel*, 116, 743–750.)

these emissions will be reduced and thermal efficiency will also increase. Crookes et al. (2006) examined the performance and emissions characteristics of spark-ignition and compression-ignition engines with simulated biogas, and they reported that the spark-ignition engine provided lower performance with biogas compared with natural gas and gasoline fuels, but this problem can be overcome by raising the compression ratio of the engine. Note, however, that NO_X would increase.

Chandra et al. (2011) studied power drop with raw biogas compared to enriched biogas. In this study, a compression-ignition engine was converted to a spark-ignition engine by decreasing its compression ratio to 12.65:1. The observed power drop was about 31.8% with CNG, 35.6 with methane-enriched biogas, and 46% with raw biogas. It can be interpreted from these results that power drop is lower—around 35%—with raw biogas than with enriched biogas.

Stoichiometric air-fuel ratio, adiabatic flame temperature, and lower heating values for different concentration levels of CO_2 in biogas were calculated, as shown in Table 6.2. Adiabatic flame temperature with fuel decreases with increasing concentrations of CO_2 in biogas. Therefore, NOx emission with raw biogas is less than base gasoline/methane.

Figure 6.4 shows the variation of NOx emission with respect to EGR and CO_2. Exhaust gas was recirculated (EGR) to the engine cylinder through the intake system during a suction stroke. The EGR percentage, which is calculated based on the ratio of EGR to total charge (EGR + air), varied up to 25%. The CO_2 in biogas was simulated up to 40%. The effect

TABLE 6.2

Properties of Various Samples of Biogas

Serial Number	% of CO_2 in Biogas	Stoichiometric A/F Ratio	Adiabatic Flame Temperature (K)	LHV_f, MJ/kg
1.	0	15.06	2324	47.44
2.	23	9.464	2252	27.40
3.	31	7.710	2207	22.56
4.	37	6.591	2168	19.45

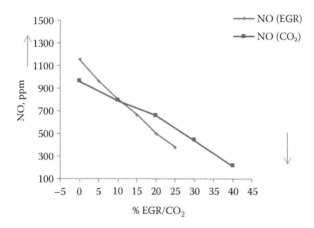

FIGURE 6.4

Variation in NO emission with EGR/CO_2 percentages. (From Bhatnagar, V. (2016), Study of performance and emissions characteristics of biogas fuelled spark-ignition engine, M.Tech. thesis, IIT Delhi.)

of EGR and CO_2 in biogas on NOx emission reduction was experimentally compared, as shown in Figures 6.5 and 6.6. The EGR mainly consists of CO_2, H_2O, N_2, and traces of CO, HC, and radicals. It is well established that EGR would reduce NOx emission significantly, but the study results show that fuel containing CO_2 could also reduce NOx emission. The CO_2 acts as a diluent and heat sink during combustion, so NOx emission will decrease with raw biogas. Therefore, the CO_2 in raw biogas has a similar effect of exhaust gas recirculation on NOx reduction, and raw biogas can also be called fuel-induced EGR. The important point is that a raw-biogas-fueled engine may not need EGR for NOx reduction, as CO_2, which is naturally present in biogas, can reduce the emission. Figure 6.5 shows the variation of NOx emission at different CO_2 content in biogas with respect to the equivalence ratio. NOx emission decreased drastically with biogas compared to compressed natural gas. However, HC emission was higher with biogas compared to compressed natural

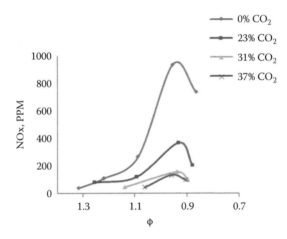

FIGURE 6.5
Variation of NOx with respect to equivalence ratio and CO_2 in biogas. (From Bhatnagar, V. (2016), Study of performance and emissions characteristics of biogas fuelled spark-ignition engine, M.Tech. thesis, IIT Delhi.)

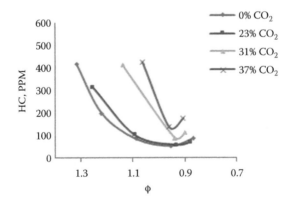

FIGURE 6.6
Variation of HC emission with respect to the equivalence ratio. (From Bhatnagar, V. (2016), Study of performance and emissions characteristics of biogas fuelled spark-ignition engine, M.Tech. thesis, IIT Delhi.)

gas, as shown in Figure 6.6. The increased HC emission is due to poor combustion mainly because of the lower flame velocity of biogas.

6.7 Utilization of Enriched Biogas in Spark-Ignition Engines

The composition of raw biogas is almost 40% carbon dioxide and only 60% natural gas, and therefore, flame velocity and calorific value are lower compared to gasoline or compressed natural gas. The combustion characteristics of biogas-fueled spark-ignition engines would be poor mainly due to the lower flame velocity of biogas, resulting higher specific fuel consumption and CO and HC emissions. Power drop with biogas is another major issue due to biogas having a lower calorific value. In order to enrich methane content in biogas, the CO_2 content in the biogas can be removed using a water-scrubbing method or membrane system. A water-scrubbing system consists of a long cylindrical column in which water at high pressure is sprayed on the top side of an inner column, whereas high-pressure biogas is passed to the bottom of the column. The directions of the water and raw biogas are opposite to each other and both fluids get mixed in the inner column. Next, the CO_2 in the biogas is dissolved in the high-pressure water stream in the column while pure methane, called enriched methane, travels to the top of the column and then the enriched biogas is trapped, compressed, and dispensed. In the case of the membrane, the desired gas is separated in the membrane, which works with differential pressure (upstream in the membrane is positive pressure (above atmosphere) and downstream is negative pressure (below atmosphere)) by membrane permeability. Membrane technology is an attractive option for a location with water scarcity.

The major advantage of using raw biogas in spark-ignition engines is that no additional energy input is required for a methane enrichment system. However, the thermal efficiency of the engine and power output are relatively lower with raw biogas. Note that OEMs may not accept this power drop in their engines/vehicles with raw biogas. Fuel transportation through a pipeline would be a problem if the raw biogas is to be injected into a natural gas grid, because both gases (raw biogas and piped natural gas (PNG)) are different compositions of natural gas. If carbon dioxide is removed from biogas, additional energy input to the enrichment is required but the thermal efficiency of the engine would improve along with an improvement in power output. A big question arises: Which fuel—raw biogas or enriched biogas—could provide net benefits? Compared to raw biogas, enriched biogas is more beneficial in terms of storage, pipeline gas transportation, better performance and emissions characteristics of spark-ignition engines, and OEM approval.

Performance and emissions characteristics of a spark-ignition engine vehicle (44.2 kW) fueled with CNG and enriched biogas (92% methane) were evaluated under the Indian driving cycle (IDC) using a chassis dynamometer (Subramanian et al., 2013). The CO, HC, and NOx emissions are slightly higher with enriched biogas than CNG but they meet the Bharat Stage (BS IV) emissions norms, as shown in Figure 6.7 (MNRE, 2014). The fuel economy of the vehicle with enriched biogas is almost equal to that with CNG. Note that the density of enriched biogas is slightly less than CNG as biogas has mainly CH_4, CO_2, and traces of H_2 and N_2, whereas CNG fuel has mainly methane (CH_4) and few percentages of ethane, propane, butane, C_5H_{15}, C_6H_{14}, and so forth. Therefore, the fuel calibration is needed for further improvement of the performance and emission characteristics of the vehicle.

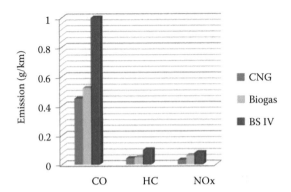

FIGURE 6.7
Emissions of a spark-ignition vehicle fueled with enriched biogas and CNG. (From MNRE (2014), Comparative evaluations of performance and mass emissions of an automotive passenger vehicle fueled with the enriched biogas using field trial tests. Project carried out by IIT Delhi by MNRE sponsorship.)

6.8 Utilization of Hydrogen in Spark-Ignition Engines

The properties of hydrogen, such as very low ignition energy (0.02 mJ), wide flammability limit (4%–75% by volume), high flame velocity (2.65 to 3.25 m/s), and high octane number are favorable as fuel for spark-ignition (IC) engines. Hydrogen has a higher octane number than gasoline fuel, so the compression ratio of spark-ignition engines can be increased, resulting in higher thermal efficiency. Hydrogen does not contain carbon, so there are no carbon-based emissions (CO, HC, CO_2, PM, etc.). On the other hand, a hydrogen-fueled engine has technical problems such as backfire, power drop, and high NOx emission.

6.8.1 Backfire

Backfire is a preignition phenomenon that occurs when a fresh hydrogen-air charge is ignited during a suction stroke. The preignition is generally initiated by localized ignition source including hot residual gas, hot spot, partially oxidized lubricating oil, unburned hydrogen, and certain radicals present in residual gas. The minimum ignition energy required for the hydrogen-air mixture is the lowest (0.02 mJ) compared to methane (0.29 mJ) and gasoline (0.24 mJ), and therefore a fresh charge could get preignited, resulting in combustion and a pressure rise in the intake manifold (Huynh et al., 2008). During backfire, the manifold pressure rose drastically up to 4 bar (Verheist et al., 2010). Many researchers, including White et al. (2006), Huynh et al. (2008), Sierens et al. (2005), and Das (1986, 2002), reported that backfire is an undesirable combustion phenomena that cannot be eliminated in carbureted hydrogen-fueled engines. Backfire could also occur when a hydrogen-air charge interacts with residual gases during the valve overlapping period (Das, 1990).

The backfire problem can be minimized with increased compression ratio, which reduces residual gas temperature (Salvi et al., 2016a). It can also be reduced by a delayed (retarded) start of a hydrogen injection during a suction stroke (Das, 1990; Mathur et al., 1991); however, it may have a negative affect by allowing residual gas to enter the intake port (Liu et al., 2008). Flame kernel growth is higher with hydrogen than with gasoline, but was reduced with exhaust gas recirculation (Salvi et al., 2016b).

6.8.2 Power Drop

Power drop is due mainly to hydrogen having a lower volumetric heat content than gasoline fuel. Mathur and Khajuria (1984) reported that a hydrogen-fueled engine can operate at a wider air-fuel ratio, resulting in an increase in thermal efficiency and a reduction in NOx emission; however, power dropped about 15% compared to base gasoline (Mathur et al., 1991). Ganesh et al. (2008) reported that a hydrogen-fueled engine provided a higher thermal efficiency of about 2%; but power dropped about 20% and NOx emission was four times higher than that of gasoline at the rated load. On the other hand, Negurescu et al. (2011) reported that an in-cylinder direct hydrogen injection system would lead to about 30% power improvement compared to a gasoline-fueled SI engine, but NOx formation increased drastically. NOx emission is significantly higher with hydrogen than with gasoline or other fuels due to higher in-cylinder temperature.

6.8.3 NOx Emission

NOx emission can be reduced in a hydrogen-fueled engine by combustion with a lean mixture, intake charge dilution (nitrogen, argon, and CO_2), water injection, and so forth. Nagalingam et al. (1983) reported that water induction with a hydrogen-air mixture in the intake manifold could significantly reduce NOx emission. Similar studies were also reported by Woolley et al. (1977) and Billings (1978) that NOx will reduce, but with the negative effect of the water in a charge possibly contaminating or degrading of the lubrication oil.

Combustion with a lean mixture strategy can reduce NOx emission in spark-ignition engines; however, the engine's power will drop (Varde et al., 1984). The charge diluents (e.g., helium, nitrogen, and water) would reduce NOx emission with the negative effect of less thermal efficiency and high power drop (Mathur et al., 1992). Aftertreatment devices, such as selective SCR devices and LNTs, could reduce NOx emission but these devices are expensive and add additional complexity to the engine (Ibrahim et al., 2008). NOx reduction at the source level is preferable in hydrogen-fueled spark-ignition engines.

Retarded spark timing will reduce NOx emission reduction in gasoline (Subramanian et al., 2009; Salvi et al., 2015) and hydrogen–natural-gas-fueled SI engines (Arroyo et al., 2015). However, this technique is not suitable for hydrogen-fueled SI engines, as it would lead to lower brake thermal efficiency (Subramanian et al., 2007). Das and Mathur (1993) reported that EGR is an effective method to control NOx emission in hydrogen-fueled spark-ignition engines. Many researchers, including Abd-Alla (2002), Subramanian et al. (2007), and Wei et al. (2012), reported that hydrogen-fueled engines could benefit from EGR (up to certain percentages) as EGR could reduce pumping work, cylinder heat losses, and oxidation of unburned fuel. However, Caton (2015) reported that in general, thermal efficiency decreased with an increasing percentage of EGR. A recent study indicates that EGR in hydrogen-fueled spark-ignition engines is more effective for NOx reduction than spark timing retarding (Salvi et al., 2016a).

Summary

The following key points, which emerged from the literature discussion, need to be considered for better performance and emissions reduction in spark-ignition engines fueled with biofuels.

Fuel Quality of Biofuels:

- The fuel quality of biofuels is desirable for improving performance and emission reduction in spark-ignition engines. The preferable biofuels for spark-ignition engines are methanol, ethanol, butanol, enriched biogas, and hydrogen, as these fuels have higher octane numbers than that of petroleum gasoline fuel. Spark-ignition engines have scope with higher octane number biofuels because the engine compression ratio of biofuels can be increased in order to obtain better thermal efficiency. In general, these biofuels are related to higher thermal efficiency of spark-ignition engines.

Alcohol (Methanol, Ethanol, and Butanol):

- Alcohol-(methanol, ethanol, and butanol) fueled engines lead to cold-start ability problems due to the fuel's higher latent heat of vaporization.
- The compression ratio of a spark-ignition engine for alcohol fuels (because of a higher octane number) has to be increased to improve the engine's thermal efficiency.
- Power of the engines would drop with alcohol fuels because they have a lower calorific value. This problem can be overcome to some extent using optimization of the compression ratio and the fuel injection system.
- CO emission with methanol, ethanol, and butanol are lower mainly due to fuel-embedded oxygen content. HC emission with these fuels is uncertain due to lower in-cylinder temperature. NOx emission is lower with methanol and ethanol; however, it is uncertain with butanol. Aldehyde emission, such as formaldehyde with methanol and acetaldehyde with ethanol, are higher.

Biogas and Enriched Biogas:

- The performance improvement and emissions reduction in a spark-ignition engine with enriched biogas is better than that of raw biogas. The main reason for poor performance with raw biogas is due to less flame velocity.

Hydrogen:

- Carbon-based emissions can be eliminated with hydrogen. The probability of backfire and higher NOx emission are disadvantages and challenges.

The technologies for utilization of gasoline in spark-ignition engines have almost matured and have been implemented in engines and vehicles, but technologies for utilization of biofuels are still in the R&D stages. The information compiled in this chapter may be useful for developing fuel-efficient, eco-friendly, dedicated biofueled spark-ignition engines.

Solved Numerical Problems

1. Calculate the adiabatic flame temperature of (i) hydrogen, (ii) methane, and (iii) gasoline at an equivalence ratio of 1.

   Input Data

   Note: Take the values of enthalpy of formation of species, specific heat, and so forth from any standard textbook or handbook of thermodynamics or combustion. The specific heat of species is taken at an average temperature of 1200 K but it has to be taken based on the corresponding temperature by iteration.

   h_{fi} = *enthalpy of formation of* H_2, O_2 *etc. at standard temperature (298 K)*

$$h_{f_{H_2}} = 0 \frac{kJ}{kmol}$$

$$h_{f_{O_2}} = 0 \frac{kJ}{kmol}$$

$$h_{f_{N_2}} = 0 \frac{kJ}{kmol}$$

$$h_{f_{CH_4}} = -74831 \frac{kJ}{kmol}$$

$$h_{f_{C_8H_{18}}} = -208447 \frac{kJ}{kmol}$$

$$T_{ref} = 298 \text{ K}$$

$$C_{P_{H_2O}} \text{ at } 1200 \text{ K} = 43.87 \frac{kJ}{kmol-K}$$

$$C_{P_{N_2}} \text{ at } 1200 \text{ K} = 33.71 \frac{kJ}{kmol-K}$$

$$C_{P_{CO_2}} \text{ at } 1200 \text{ K} = 56.21 \frac{kJ}{kmol-K}$$

   Adiabatic flame temperature of hydrogen, methane and gasoline at an equivalence ratio of one = ?

Solution

(i) Adiabatic Flame Temperature for Hydrogen

Stoichiometric equation for hydrogen fuel:

$$H_2 + a(O_2 + 3.76\,N_2) \rightarrow b\,H_2O + a\,3.76\,N_2$$

$$H_2 + 0.5(O_2 + 3.76\,N_2) \rightarrow H_2O + 1.88\,N_2$$

Calculations of adiabatic flame temperature:

Enthalpy of reactants (H_R) = enthalpy of products (H_P)

$H_R = H_P$

H = enthalpy of formation (h_f) + sensible enthalpy $(\bar{C}_p\,dT)$

Enthalpy of reactants can be written as follows:

$$H_R = h_{f_{H_2}} + 0.5\,h_{f_{O_2}} + 1.88\,h_{f_{N_2}} = 0 + 0 + 0 = 0$$

Enthalpy of products can be written as given below:

$$H_P = h_{f_{H_2O}} + C_{P_{H_2O}}(T_{ad} - T_{ref}) + 1.88\,h_{f_{N_2}} + 1.88\,C_{P_{N_2}}(T_{ad} - T_{ref})$$

$$H_R = H_P$$

$$0 = (-241845) + 43.87\,(T_{ad} - 298) + 0 + 1.88 \times 33.71\,(T_{ad} - 298)$$

$$T_{ad} = 2553\,K$$

(ii) Adiabatic Flame Temperature for Methane

Stoichiometric equation for methane

$$CH_4 + a(O_2 + 3.76\,N_2) \rightarrow b\,H_2O + a\,3.76\,N_2 + c\,CO_2$$

$$CH_4 + 2(O_2 + 3.76\,N_2) \rightarrow 2\,H_2O + 7.52\,N_2 + CO_2$$

$$\textit{Enthalpy of reactants, } H_R = h_{f_{CH_4}} + 2\,h_{f_{O_2}} + 7.52\,h_{f_{N_2}}$$
$$= -74831 + 0 + 0$$
$$= -74831\ \text{kJ/kmol}$$

Enthalpy formation at product side is

$$\textit{Enthalpy of products, } H_P = 2\left[h_{f_{H_2O}} + C_{P_{H_2O}}(T_{ad} - T_{ref})\right]$$
$$+ 7.52\left[h_{f_{N_2}} + C_{P_{N_2}}(T_{ad} - T_{ref})\right] + \left[h_{f_{CO2}} + C_{P_{CO2}}(T_{ad} - T_{ref})\right]$$

$$H_R = H_P$$

$$-74831 = 2[-241845 + 43.87\,(T_{ad} - 298)] + 7.52 \times 33.71\,(T_{ad} - 298) - 393546$$
$$+ 56.21\,(T_{ad} - 298)$$

$$T_{ad} = 2316.88\ \textbf{K}$$

(iii) Adiabatic Flame Temperature for Gasoline

Stoichiometric equation for gasoline:

$$C_8H_{18} + a(O_2 + 3.76\,N_2) \rightarrow bH_2O + a3.76\,N_2 + CO_2$$

$$C_8H_{18} + 12.5\,(O_2 + 3.76\,N_2) \rightarrow 9\,H_2O + 47\,N_2 + 8\,CO_2$$

Enthalpy of reactants, $H_R = h_{f_{C8H18}} + 12.5\,h_{f_{O_2}} + 47\,h_{f_{N_2}} = -208447$ kJ/kmol

Enthalpy of products, $H_P = 9\left[h_{f_{H_2O}} + C_{P_{H_2O}}(T_{ad} - T_{ref}) \right]$
$$+ 47\,C_{P_{N_2}}(T_{ad} - T_{ref}) + 8\left[h_{f_{CO2}} + C_{P_{CO2}}(T_{ad} - T_{ref}) \right]$$

$$H_R = H_P$$

$$-208447 = 9[-241845 + 43.87\,(T_{ad} - 298)]$$
$$+ 47 \times 33.71\,(T_{ad} - 298) + 8\,[-393546] + 56.21\,(T_{ad} - 298)]$$

$$T_{ad} = \textbf{2404.6 K}$$

2. Calculate adiabatic flame temperature of hydrogen with 10% EGR (by volume) with exhaust gas temperature of 350 K and also percentage of thermal dilution.

Stoichiometric Equation for hydrogen fuel:

$$H_2 + a(O_2 + 3.76\,N_2) \rightarrow bH_2O + a3.76\,N_2$$

$$H_2 + 0.5\,(O_2 + 3.76\,N_2) \rightarrow H_2O + 1.88\,N_2$$

$$H_R = H_P$$

$$\Sigma_{react}\, N_i h_i = \Sigma_{prod}\, N_i h_i$$

where N_i is the number of moles of i species

$$\text{Mole fraction of exhaust gases} = x_{H_2O} = \frac{1}{1+1.88} = 0.3472$$

$$x_{N_2} = \frac{1.88}{1+1.88} = 0.6527$$

Molar-specific enthalpy of exhaust gases at 350 K (h_{EGR}) = 0.3472(–242001) + 0.6527 (0) = –84022.74 kJ/kmol

$N_{H2} = 1$, $N_{AIR} = 2.38$, $N_{EGR} = (N_F + N_A) \times (\%EGR)/100$

$H_R = N_F h_F + N_A h_A + N_{EGR} h_{EGR}$

For 10% EGR

$$H_R = 1\,(0) + 2.38\,(0) + 3.38(-84022.74)\left(\frac{10}{100}\right) = -28399.68\ kJ/kmol$$

$$H_P = h_{f_{H_2O}} + C_{P_{H_2O}}(T_{ad} - 298) + 1.88 C_{P_{N_2}}(T_{ad} - 298)$$

$$= (-241845) + 43.87\,(T_{ad} - 298) + 1.88 \times 33.71\,(T_{ad} - 298)$$

$$= -273803.95 + 107.27\,T_{ad}$$

$$H_R = H_P$$

$$T_{ad} = 2287.5\ K$$

Thermal Dilution (TD)

$$TD = \frac{T_{ad} - T_{ad-EGR}}{T_{ad}} \times 100 = \frac{2553 - 2287.5}{2553} \times 100 = 10.39\%$$

3. Calculate the adiabatic flame temperature of hydrogen with water injection (10%) and also the thermal dilution. Take the equivalence ratio of the mixture as one.

   Stoichiometric equation for hydrogen fuel:

   $$H_2 + a(O_2 + 3.76 N_2) + c\,H_2O \rightarrow b\,H_2O + a\,3.76 N_2$$

   $$H_2 + 0.5(O_2 + 3.76 N_2) + H_2O \rightarrow 2H_2O + 1.88 N_2$$

For (W/F = 2.25/1 mass basis) 0.25:1 mole basis

$$H_R = N_F h_F + N_A h_A + N_W h_W$$

$$H_R = 1(0) + 2.38(0) + 0.25(-241845) = -60461.25$$

$$H_P = 1.25[h_{f_{H_2O}} + C_{P_{H_2O}}(T_{ad} - 298)] + 1.88 C_{P_{N_2}}(T_{ad} - 298)$$

$$1.5[(-241845) + 43.87(T_{ad} - 298)] + 1.88 \times 33.71(T_{ad} - 298)$$

$$T_{ad} = 2343.9 \, K$$

For (W/F = 4.5/1 mass basis) 0.5:1 mole basis

$$H_R = N_F h_F + N_A h_A + N_W h_W$$

$$H_R = 1(0) + 2.38(0) + 0.5(-241845) = -120922.5$$

$$H_P = 1.5[h_{f_{H_2O}} + C_{P_{H_2O}}(T_{ad} - 298)] + 1.88 C_{P_{N_2}}(T_{ad} - 298)$$

$$= 1.5[(-241845) + 43.87(T_{ad} - 298)] + 1.88 \times 33.71(T_{ad} - 298)$$

$$T_{ad} = 2170.15 \, K$$

For (W/F = 9/1 mass basis) 1:1 mole basis

$$H_R = N_F h_F + N_A h_A + N_W h_W$$

$$H_R = 1(0) + 2.38(0) + 1(-241845) = -241845$$

$$H_P = 2[h_{f_{H_2O}} + C_{P_{H_2O}}(T_{ad} - 298)] + 1.88 C_{P_{N_2}}(T_{ad} - 298)$$

$$= 2[(-241845) + 43.87(T_{ad} - 298)] + 1.88 \times 33.71(T_{ad} - 298)$$

$$T_{ad} = 1898.46 \, K$$

Thermal Dilution (TD):

$$TD = \frac{T_{ad} - T_{ad-EGR}}{T_{ad}} \times 100 = \frac{2553 - 2343.9}{2553} \times 100 = 8.19\%$$

Similarly, the percentage of thermal dilution for different water-to-fuel ratios are calculated as given in the following table.

W/F Ratio	Adiabatic Temperature T_{ad} (K)	Thermal Dilution (%)
0:1	2553	–
2.25:1	2343.9	8.19
4.5:1	2170.15	15
9:1	1898.46	25.63

4. A methane-fueled spark-ignition engine with a compression ratio of 9:1 operates at an equivalence ratio of 0.8. If the compression ratio is changed from 9:1 to 11:1, calculate the percentage of change in the flame velocity of methane.

Input Data

Compression Ratio = 0.8:1

Percentage of change in flame velocity of methane = L

Solution

Given data: $\varphi = 0.8$, $T_0 = 298$ K, $P_0 = 1$ bar

$$S_L = S_{L,0} \left(\frac{T_u}{T_0}\right)^\alpha \left(\frac{P}{P_0}\right)^\beta$$

$$\alpha = 2.18 - 0.8(\varnothing - 1)$$

$$\beta = -0.16 + 0.22(\varnothing - 1)$$

$\alpha = 2.02$

$\beta = -0.204$

$S_{LO} = 29$ cm/s (Heywood, 1988)

$r_c = 9$

$(r_c)^{\Upsilon-1} = (P_2/P_1)^{\Upsilon-1/\Upsilon} = (T_2/T_1)$ [$\Upsilon = 1.4$]

$P_0 = P_1 = 1$ bar; $T_0 = T_1 = 298$ K

For $r_c = 9:1$, $T_u = T_2 = 717.65$ K; $P = P_2 = 21.84$ bar

$S_L = 91.24$ cm/s

$r_c = 11$

$T_u = T_2 = 777.63$ K; $P = P_2 = 28.94$ bar

$S_L = 101.32$ cm/s

Percentage increase in flame velocity = 101.32 – 91.24/91.24 = 0.11

Flame velocity is increased by 11%.

5. The gap between electrodes in a spark plug in an ethanol-fueled spark-ignition engine is 0.5 mm. Calculate the minimum ignition energy requirement for initiation of ignition of the charge trapped between the electrodes.

Solution

Input parameters:

$D = 0.5$ mm

Assumption:

Density of ethanol at 300 K, $d = 0.796 \times 10^3$ kg/m^3

Specific heat at constant pressure, $C_p = 2.45$ kJ/kg K

Electrode gap, $D = 0.5$ mm

Flame temperature, $T_f = 2193$ K

Reactant temperature $(T_r) = 300$ K

Minimum ignition energy (MIE), mJ = ?

$$MIE \text{ (mJ)} = dC_p\pi D^3(T_f - T_r) \times 1000/6$$
$$MIE = \left[0.796 \times 10^3 \times 2.45 \times 10^3 \times \pi \times (0.5 \times 10^{-3})^3\right.$$
$$\left. \times (2193 - 300)\right] \times 1000/6 = 241.62 \text{ mJ}$$

6. The electrode gap in a spark plug in a hydrogen-fueled spark-ignition engine is 1 mm. Calculate the minimum ignition energy requirement for initiation of ignition of the charge trapped between the electrodes.

Solution

Input parameters:

$D = 1$ mm

Assumptions:

Density of hydrogen at 300 K, $d = 0.08078$ kg/m^3

Specific heat at constant pressure, $C_p = 14.31$ kJ/kg K

Electrode gap, $D = 1$ mm

Flame temperature, $T_f = 3073$ K

Reactant temperature $(T_r) = 300$ K

Minimum ignition energy (MIE), mJ = ?

$$MIE \text{ (mJ)} = dC_p\pi D^3(T_f - T_r) \times 1000/6$$
$$MIE = \left[0.08078 \times 14.31 \times 10^3 \times \pi \times (1 \times 10^{-3})^3\right.$$
$$\left. \times (3073 - 300)\right] \times 1000/6 = 1.68 \text{ mJ}$$

7. Calculate the mass fraction burn rate of a gasoline-fueled spark-ignition engine using the Wiebe function. The start and end of combustion are 2 ^{0}btdc and 40 ^{0}atdc.

Solution

Input parameters:

$\Delta\theta = 42^0$ (duration of combustion)

$\theta_0 = 2$ btdc (start of combustion)

$\theta =$ Crank angle at which mass burned has to be calculated

Mass burned fraction

$$X_b = 1 - \exp\left[-a\left(\frac{\theta - \theta_0}{\Delta\theta}\right)^{m+1}\right]$$

$a = 5, m = 2$

CA	X_b	CA	X_b
358	0	381	0.560061
359	6.75E-05	382	0.606606
360	0.00054	383	0.65163
361	0.00182	384	0.694607
362	0.00431	385	0.735087
363	0.0084	386	0.772699
364	0.014472	387	0.80717
365	0.022882	388	0.838323
366	0.033963	389	0.86608
367	0.048008	390	0.890455
368	0.06526	391	0.911548
369	0.085909	392	0.929527
370	0.110075	393	0.944619
371	0.137801	394	0.957092
372	0.16905	395	0.967236
373	0.203692	396	0.975355
374	0.241512	397	0.981744
375	0.282201	398	0.986689
376	0.325367	399	0.990451
377	0.370542	400	0.993262
378	0.417193		
379	0.464739		
380	0.512568		

8. A methanol-fueled spark-ignition engine operates at an equivalence ratio of 0.9. The mass flow rate and calorific value of methanol fuel are 10 kg/hr and 23,000 kJ/kg respectively. The brake thermal efficiency of the engine is 35%. The blow by loss is 10%. The engine emits 0.2% (volume) of CO, 0.1% (volume) of UHC, and 750 ppm of NOx. Calculate the specific mass emissions.

Solution

Input Parameters:

$\phi = 0.9$

Mass flow rate of fuel, $m_f = 10$ kg/hr

Calorific value of methanol, $CV_{methanol} = 23000$ kJ/kg

Solution:

Stoichiometric equation for combustion of methanol is as given below:

$CH_3OH + 1.5(O_2 + 3.77N_2) = CO_2 + 2H_2O + 5.66N_2$

Stoichiometric air-fuel ratio ($\phi = 1$), $(A/F)_s = 6.47$; $(A/F)_s = 1/6.47 = 0.154$

$\phi = (A/F)_a/(A/F)_s = (A/F)_a/0.154 = 0.9$

$(A/F)_a = 0.9 \times 0.154 = 0.139$

Mass flow rate of air, m_a = mass flow rate of fuel/0.139 = 10/0.139 = 71.94 kg/hr

Mass flow rate of exhaust, $m_{exh.} = (m_f + m_a) \times 0.9$ (10% is blowby loss)

$m_{exh.} = (10 + 71.94) \times 0.9 = 73.75$ kg/hr

Power output of the spark-ignition engine, $P = \eta \times m_f \times CV_{methanol}$

$= (0.35 \times 10 \times 23000 \times 1000)/3600$

$= 22.36$ kW

Specific mass emissions can be calculated using the following equation:

$$\dot{m}_{e,i} = \frac{x_i \times MW_i \times \dot{m}_{ex} \times 3600}{MW_{ex} \times BP}$$

where

$\dot{m}_{e,i}$ (kg/kW-hr) = specific mass emission of species i (i = CO, HC, NOx, CO_2, CH_4, N_2O, etc.)

x_i = zvolume fraction of emission i (i = %/100 (or) ppm $\times 10^{-6}$)

MW_i = molecular weight of emission i

$\dot{m}_{ex}$ = mass flow rate of exhaust gas (kg/s)

MW_{ex} = molecular weight of exhaust gas (= 28.6 g/mol)

Thus, specific emission of CO $= \dfrac{(0.2 \times 28 \times 73.75 \times 1000)}{(100 \times 28.6 \times 22.36)} = 6.46$ g/kW-hr

Specific emission of UHC $= \dfrac{(0.1 \times 16 \times 73.75 \times 1000)}{(100 \times 28.6 \times 22.36)} = 1.85$ g/kW-hr

Specific emission of NOx $= \dfrac{(750 \times 10^{-6} \times 30 \times 73.75 \times 1000)}{(28.6 \times 22.36)} = 2.59$ g/kW-hr

9. The composition of raw biogas is 60% methane (by weight), 38% CO_2 (by weight), 1.5% N_2, 500 ppm H_2S, and the remainder is hydrogen. The flow rate of biogas in a spark-ignition engine is 2 kg/hr. If combustion is perfect, calculate the mass flow rate of CO_2 emission from the engine.

Solution

Input parameters:

Mass flow of raw biogas in one hour = 2 kg

CO_2 is 38% (by weight) in biogas. Therefore, in 2 kg of biogas, the amount of CO_2 will be

$$2 \times 0.38 = 0.76 \text{ kg/hr}$$

Methane in biogas undergoes perfect combustion according to the following equation:

$$CH_4 + 2O_2 = CO_2 + 2H_2O$$

16 gm CH_4 produces 44 gm CO_2

1.2 kg CH_4 produces = (44 × 1.2)/16 CO_2 per hour = 3.3 kg/hr

Total mass flow rate of CO_2 emission = 3.3 + 0.76 = 4.06 kg/hr

10. Calculate the heat of combustion for hydrogen in a standard condition.

Solution

Combustion reaction of hydrogen:

$$H_2 + \frac{1}{2}O_2 = H_2O(g)$$

Standard heat of formation, $\Delta H_f(H_2)$ = 0 MJ/kmol

$\Delta H_f(O_2)$ = 0 MJ/kmol, $\Delta H_f(H_2O(g))$ = –241.83 MJ/kmol

Heat of combustion, $\Delta H_c = \Delta H_f$ (products) – ΔH_f(reactants)

ΔH_c = –241.83 – (0 + 0) = –241.83 MJ/kmol

11. An industrial company produces 5000 SCM landfill gas/day. The gas composition is 60% CH_4 and 40% CO_2 (by volume). The landfill gas, CO and HC heat content are 28.8, 15 and 44 MJ/kg, respectively. Calculate the following.

   1. Maximum power output of a spark-ignition engine fueled with raw biogas. The industry plans to operate a gas generator with raw biogas for 6 hours/day. Thermal efficiency: 32%, *BMEP:* 14 bar, speed: 1500 rpm, methane's calorific value: 48 MJ/kg.

   2. Number of cylinders for the spark-ignition engine.

   3. Number of vehicles to be fueled with enriched landfill gas. A 5-tonne capacity truck with 4 km/liter fuel economy travels (to and from): 200 km + 200 km. The energy input to an enriched biogas plant is 30% of raw biogas input.

4. Combustion efficiency if the engine emits 2 g/s of CO and 4 g/s of HC.

5. CO_2 emissions from the spark-ignition engine.

Solution

Input parameters:

Available gas = 5000 m³

Thermal efficiency = 32%

$BMEP$ = 14 bar = 14 × 10⁵ N/m²

Calorific value = 48 MJ/kg

Assumptions:

Bore × stroke of cylinder = 100 mm

Density of methane = 0.6 kg/m³

Density of carbon dioxide = 1.98 kg/m³

Calorific value of CO_2 = 0

Raw gas is enriched to 90% methane.

1. Power output of engine, $P = \eta \times \dot{m}_f \times CV$

Calorific value of landfill gas, CV = 28.8 MJ/kg

Density of landfill gas = 0.6 × .65 + 0.4 × 1.98 = 1.182 kg/m³

Mass of landfill gas = 5000 × 1.182 = 5910 kg

Mass flow rate, m_f = 5910/6 × 3600 = 0.273 kg/s

P = 0.32 × 0.273 × 28.8 × MW (η = 0.32 given)

P = 2.515 MW

2. Power output of engine,

$$P = BMEP \times V_s \times \frac{N}{2} \times K$$

where V_s = swept volume of cylinder= π × 100² × 100 × 10⁻⁹ m³

K = Number of cylinders

2.515 × 10⁶ = 14 × 10⁵ × 0.00785 × (1500/2 × 60) × K

K = 183 (as a single genset with 183 cylinders for this engine's specification may not be available in the market/manufacturers; for example, 6 numbers of genset (each 420 kW) with 32 cylinder may be selected to produce the required power or genset with the specified number of cylinders available with the manufacturer may be selected for this desired power generation).

3. If 30% raw gas is used in an enrichment process then it is not available for utilization.

5000 × 30% = 1500 m³ is not available.

5000 – 1500 = 3500 m³ is the available raw gas. Methane is 60% of 3500 = 2100 m³

If landfill gas is enriched to 90% methane, the volume of landfill gas 2100 × 100/90 = 2333.33 m³

One truck uses 1 liter of gas for 4 km

For 400 km, 100 liters of gas is required, which is equivalent to 0.1 m³

Next, 0.1 m³ is used by 1 truck, and then 2333.33 m³ will be used by 2333.33/0.1 = 23333 trucks

4. Combustion efficiency is given by

$$\eta = 1 - \frac{(m_a + m_f)\sum x_{iCV_i}}{m_f \times CV_f}$$

Mass flow rate of CO = 2 g/s
Mass flow rate of HC = 4 g/s

  If raw gas is used:

$$\eta = 1 - (2 \times 15 \times 10^{-3} + 4 \times 44 \times 10^{-3})/28.8 \times 0.273 = 97.37\%$$

  If enriched gas is used:

$$\eta = 1 - (2 \times 15 \times 10^{-3} + 4.44 \times 10^{-3})/43.2 \times 0.273 = 98.25\%$$

12. An Otto cycle engine operates with a compression ratio of 10:1. The mass flow rate of gasoline fuel (C_8H_{18}) is 10 kg/hr. Calculate the thermal efficiency and CO_2 emission of the engine. Let the engine operate with a stoichiometric air-fuel ratio and complete combustion.

Solution

Input parameters:

Compression ratio, r_c = 10

Fuel flow rate, 10 kg/hr

Assumptions:

Ratio of specific heats, Υ = 1.4

Thermal efficiency of air standard Otto cycle is given below:

$$\eta = 1 - \left(\frac{1}{r_c}\right)^{\gamma - 1}$$

$$\eta = 1 - 1/10^{0.4} \times 100 = 60.8\%$$

Combustion equation of gasoline is given as:

$$C_8H_{18} + 12.5O_2 = 8CO_2 + 9H_2O$$

114 gm of gasoline produces 352 gm of carbon dioxide

10 kg of gasoline per hour will produce 352 × 10/114 kg/hr of CO_2 = 30.87 kg/hr

13. A chloro-alkali industrial company produces 50,000 SCM of hydrogen as a tangible product. The total required electricity demand for the industry for 16 hours (two shifts) is 50 MW, which is supplied by a natural-gas-fueled power plant. The industrial company power demand is increased by 10% of total power due to a pilot-scale unit operation. The industrial company plans to produce additional power of 5 MW (10% of total power) using and internal combustion engine fueled with hydrogen. In addition, the industrial company uses 50 internal combustion engine trucks for transporting the finished products (chlorine, sulfuric acid, etc.) from the industrial company to the consumer. The trucks are fueled with compressed natural gas with a fuel economy of 5 km/kg of CNG and travel 300 km back and forth per day. The thermal efficiency of the power plant and trucks are 35%. The calorific value of CNG and hydrogen are 43 MJ/kg and 120 MJ/kg, respectively. The density of hydrogen and CNG are 0.07 kg/m³ and 0.8 kg/m³, respectively. Calculate the following parameters:

1. Whether the available hydrogen is sufficient for both generation of the required supplementary power and fuel for the trucks.
2. Whether a surplus quantity of hydrogen is available (after utilizing it for power generation and transportation).
3. Quantity of CO_2 emission reduction with hydrogen for both power generation and fuel for 50 trucks compared to the base CNG.

Solution

Input parameters:

Available hydrogen = 50,000 SCM

Hydrogen is required for 16 hours for power generation and 12 hours for transportation.

Thermal efficiency of power plant and trucks = 35%

CV of hydrogen = 120 MJ/kg

CV of CNG = 43 MJ/kg

Total distance traveled by one truck per day = 300 km

Total number of trucks = 50

Assumption:

CNG is assumed to have 100% methane and undergoes complete combustion.

   1. If only hydrogen is used for power generation:

      Mass of hydrogen = 50000 × 0.07 = 3500 kg

      Mass flow rate = 3500/16 × 3600 = 0.0607 kg/s

      Power = 0.35 × 0.0607 × 120 = 2.55 MW while demand is 5 MW.

      Power demand by transport sector: (using CNG)

      Total requirement of CNG by 50 trucks = 300 × 50/5 = 3000 kg

      Mass flow rate = 3000/12 × 3600 = 0.069 kg/s

      Power = 0.35 × 0.069 × 43 = 1.04 MW

      Total demand = 5 + 1.04 = 6.04 MW, which cannot be met by hydrogen alone

   2. There is no surplus quantity of hydrogen

   3. CNG combustion equation:

$$CH_4 + 2O_2 = CO_2 + 2H_2O$$

For transportation, 3000 kg CNG is required

$$CO_2 \text{ produced by CNG} = 44 \times 3000/16 = 8250 \text{ kg}$$

Mass of CNG required for 55 MW power:

$$m = 55 \times 10^6 \times 16 \times 3600/0.35 \times 43 \times 10^6 = 210498.33 \text{ kg}$$

It will produce CO_2 = 44 × 210498.33/16 = 578870.43 kg

Therefore, the total amount of CO_2 produced in both sectors = 578870.43 + 8250 = 587120.43 kg

14. A propane-fueled spark-ignition engine operates with 20% excess air. Calculate the mole fraction of oxygen in the exhaust gas.

Solution

Parameter:

20% excess air

Stoichiometric combustion equation of propane:

$$C_3H_8 + 5(O_2 + 3.76N_2) = 3CO_2 + 4H_2O + 18.8N_2$$

With 20% excess air, oxygen is also present in exhaust.

$$C_3H_8 + 1.2 \times 5(O_2 + 3.76N_2) = 3CO_2 + 4H_2O + O_2 + 22.56N_2$$

Mole fraction of oxygen in exhaust = $1/(3 + 4 + 1 + 22.56) = 0.0376$

15. The fuel and air flow rates of a gasoline-fueled automotive spark-ignition engine are 8 g/sec and 160 g/sec, respectively. The exhaust gas contains 2% CO, 1% HC, and 0.5% soot. Take the calorific value of CO, HC, and soot as 10.1, 42, and 32.8 MJ/kg, respectively. Calculate its combustion efficiency.

Solution

Parameter:

X_i = 2%, 1%, 0.5%

CV_i = 10.1 MJ/kg, 42 MJ/kg, 32.8 MJ/kg

m_f = 8 g/s, CV_f = 43 MJ/kg

m_a = 160 g/s

Combustion efficiency is given by

$$\eta = 1 - \frac{(m_a + m_f)\sum x_{iCV_i}}{m_f \times CV_f}$$

$$\eta = 1 - (160 + 8)(0.02 \times 10.1 + 0.01 \times 42$$
$$+ .005 \times 32.8)/0.008 \times 43 = 0.62 \text{ or } 62\%$$

16. An engine operates with gasoline as fuel (CV: 44 MJ/kg). If the engine operates with lower calorific ethanol fuel (CV: 28 MJ/kg), calculate the power drop in the engine. Assume the injection system in the engine can supply 10% excess fuel.

    Solution

    Assume thermal efficiency in both cases are the same

    Power output in the first case, $P1 = \eta \times \dot{m}_f \times CV = \eta \times m \times 44$

    Power output in the second case, $P2 = \eta \times (1.1 \ m) \times 28$

    $P1/P2 = 44/(1.1 \times 28) = 1.42$

    $P1 = 1.42P2$

    Power drop $= \{(1.42P2 - P2)/1.42P2\}100\% = 29.57\%$

17. An ethanol-gasoline (E15) fueled engine emits CO_2 emission of 2 kg/kg of E15. Calculate the percentage of CO_2 and the percentage of CO_2 to be reduced as a carbon-neutral fuel.

    Solution

    Input Parameters:

    Density of ethanol = 740 kg/m³

    Density of gasoline = 780 kg/m³

    Calorific value of ethanol = 29.7 MJ/kg

    Calorific value of gasoline = 47.3 MJ/kg

    Per kg fuel burned produces 2 kg CO_2

    Assumptions:

    Efficiency of engine = 30%

    Brake power = 10 kW

    Time duration of operation of engine = 5 hours

    Carbon dioxide produced by ethanol is fixed by feedstock such as sugarcane.

    Power in an engine is given by the following equation:

    $$P = \eta \times \dot{m}_f \times CV$$

    $$CV = 0.15 \times 740 \times 29.7 + 0.85 \times 780 \times 47.3 = 34.656 \text{ MJ/kg}$$

    $$10 \times 1000 = 0.3 \times m_f \times 34.656 \times 10^6$$

$$m_f = 0.09618 \, \text{mg/sec}$$

In 5 hours $0.744 \times 10^{-3} \times 5 \times 3600 = 1.731$ kg of fuel is consumed.

$$\text{Total CO}_2 \text{ produced} = 1.731 \times 2 = 3.462 \text{ kg}$$

Combustion equation of gasoline:

$$C_8H_{18} + 12.5(O_2 + 3.76 N_2) = 8 CO_2 + 9 H_2O$$

$$\text{Per kg gasoline CO}_2 \text{ emission} = 8 \times 44/114 = 3.08 \text{ kg}$$

By using E15, CO_2 reduction:

$$\text{Energy rate of ethanol} = 0.15 \times 740 \times 29.7 = 3296.7 \text{ MJ}$$

$$\text{Energy rate of gasoline} = 0.85 \times 780 \times 47.3 = 31359.9 \text{ MJ}$$

$$\text{Total energy rate of blend} = 3296.7 + 31359.9 = 34656.6 \text{ MJ}$$

$$\text{Energy share of ethanol} = 3296.7/34656.6 = 9.5\%$$

Therefore, by using E15 blend = 9.5% CO_2 emission could be prevented.

$$9.5 \times 3.08/100 = 0.293 \, \text{kg CO}_2 \text{ per kg of blend.}$$

18. Calculate the instantaneous NO emission of a hydrogen-fueled spark-ignition engine if the engine operates at an equivalence ratio of 0.8. The adiabatic flame temperature of hydrogen at an equivalence ratio (ϕ) of 0.8 is 2208 K.

    Input data

    The adiabatic flame temperature of hydrogen at an equivalence ratio (ϕ) of 0.8, $T = 2208 \, K$

    Instantaneous NO emission = ?

Solution

The combustion equation of hydrogen for $\phi = 0.8$ is given below:

$H_2 + 0.62(O_2 + 3.76\ N_2) = H_2O + 2.33N_2 + 0.12O_2$

Mole fraction of N_2, $X_{N_2} = 2.33/(1 + 2.33 + 0.12) = 0.675$

Mole fraction of O_2, $X_{O_2} = 0.12/(1 + 2.33 + 0.12) = 0.035$

Converting mole fractions into molar concentrations,

$[N_2] = X_{N_2}\ P/(R_u T)$ (where, P is the atmospheric pressure and R_u is the universal gas constant)

$[N_2] = 0.675 \times 101325/(8315 \times 2208) = 3.72 \times 10^{-3}$ kmol/m³ $= 3.72 \times 10^{-6}$ mol/cm³

$[O_2] = X_{O_2}\ P/(R_u T)$

$[O_2] = 0.035 \times 101325/(8315 \times 2208) = 1.93 \times 10^{-4}$ kmol/m³ $= 1.67 \times 10^{-7}$ mol/cm³

Putting these values in the following equation (Heywood, 1988)

$$\frac{d[NO]}{dt} = \frac{6 \times 10^{16}}{T^{1/2}} \exp\left(-\frac{69090}{T}\right)[O_2]_e^{1/2}\left[N_2\right]_e$$

$$\frac{d[NO]}{dt} = 4.99 \times 10^{-8}\ \text{mol/cm}^3\text{-s} = 4.99 \times 10^{-5}\ \text{kmol/m}^3\text{-s}$$

In terms of ppm

$$\frac{d(X_{NO})}{dt} = \frac{R_u T}{P}\frac{d[NO]}{dt}$$

$$\frac{d(X_{NO})}{dt} = \frac{8315 \times 2208}{101325} \times 4.99 \times 10^{-5} = 9.053 \times 10^{-3}\ (\text{kmol/kmol})/\text{s} = 9053\ \text{ppm/s}$$

Inference: The calculated value of NO is 9053 ppm/s, which is higher than the actual NO emission formed in an engine because the dilution ratio, chemical reaction of N/N2 with O, O_2, and OH with other residual and intermediate products, varying in in-cylinder pressure and temperature, reaction time, and so forth, are not considered for this numerical problem. If these parameters are considered to calculate NO emission, the actual NO emission value will be much lower than the theoretically calculated value.

19. A spark-ignition engine burns a rich isooctane mixture. Exhaust gas analysis on a dry volumetric basis gives the following results:

$CO_2 = 9.8\%$

$CO = 4.5\%$

$CH_4 = 2\%$

$H_2 = 0.85\%$

$O_2 = 2.8\%$

$N_2 = 80.05\%$

Calculate the fuel-air equivalence ratio and the air-fuel ratio on a mass basis.

Solution

Stoichiometric equation:

$$C_8H_{18} + 12.5(O_2 + 3.76N_2) = 8CO_2 + 9H_2O + 47N_2$$

Actual equation:

$$\phi C_8H_{18} + 12.5(O_2 + 3.76N_2) = a[0.098\,CO_2 + 0.045\,CO + 0.02\,CH_4$$
$$+ 0.0085\,H_2 + 0.028\,O_2 + 0.80\,N_2] + b\,H_2O$$

Nitrogen balance:

$$12.5 \times 3.76 = a \times 0.8$$

$a = 58.75$

Carbon balance:

$$8\phi = (0.098 + 0.045 + 0.02)58.75$$

$\phi = 1.19$

Oxygen balance:

$$2 \times 12.5 = (2 \times a \times 0.098 + a \times 0.045 + 2 \times a \times 0.028) + b$$

$b = 7.551$

Air-fuel ratio by mass $= 12.5 \times 4.765 \times 28.97/1.19 \times 114 = 12.7$

References

Abd-Alla, G. H. (2002), Using exhaust gas recirculation in internal combustion engines: A review. *Energy Conversion and Management*, 43, 1027–1042.

Al-Hasan, M. (2003), Effect of ethanol–unleaded gasoline blends on engine performance and exhaust emission, *Energy Conversion and Management*, 44(9), 1547–1561.

Amer, A., Babiker, H., Chang, J., Kalghatgi, G. et al. (2012), Fuel effects on knock in a highly boosted direct injection spark-ignition engine, *SAE International Journal of Fuels and Lubricants*, 5, 1048–1065.

Arroyo, J., Moreno, F., Muñoz, M. and Monné, C. (2015), Experimental study of ignition timing and supercharging effects on a gasoline engine fueled with synthetic gases extracted from biogas, *Energy Conversion and Management*, 97, 196–211.

Bhatnagar, V. (2016), Study of performance and emissions characteristics of biogas fuelled spark-ignition engine, M.Tech. thesis, IIT Delhi.

Billings, R. E. (1978), Hydrogen-powered mass transit system, *International Journal of Hydrogen Energy*, 3, 49–59.

Broustail, G., Halter, F., Seers, P., Moréac, G. et al. (2012), Comparison of regulated and non-regulated pollutants with iso-octane/butanol and iso-octane/ethanol blends in a port-fuel injection spark-ignition engine, *Fuel*, 94, 251–261.

Caton, J. A. (2015), A thermodynamic comparison of external and internal exhaust gas dilution for high-efficiency internal combustion engines, *International Journal of Engine Research*, 1–21.

Celik, M. B. (2008), Experimental determination of suitable ethanol–gasoline blend rate at high compression ratio for gasoline engine, *Applied Thermal Engineering*, 28(5–6), 396–404.

Çelik, M. B., Özdalyan, B. and Alkan, F. (2011), The use of pure methanol as fuel at high compression ratio in a single cylinder gasoline engine, *Fuel*, 90(4), 1591–1598.

Chandra, R., Vijay, V. K., Subbarao, P. M. V. and Khura, T. K. (2011), Performance evaluation of a constant speed IC engine on CNG, methane enriched biogas and biogas, *Applied Energy*, 88, 3969–3977.

Crookes, R. J. (2006), Comparative bio-fuel performance in internal combustion engine, *Biomass and Bioenergy*, 30, 461–468.

Das, L. M. (1986), Studies on timed manifold injection in hydrogen operated spark-ignition engine: Performance, combustion and exhaust emission characteristics, PhD thesis, Indian Institute of Technology Delhi.

Das, L. M. (1990), Fuel induction techniques for a hydrogen operated engine, *International Journal of Hydrogen Energy*, 15(11), 833–842.

Das, L. M. (2002), Hydrogen engine: Research and development (R&D) programmes in Indian Institute of Technology (IIT), Delhi, *International Journal of Hydrogen Energy*, 27, 953–965.

Das, L. M. and Mathur, R. (1993), Exhaust gas recirculation for NOx control in a multi cylinder hydrogen-supplemented S.I. engine, *International Journal of Hydrogen Energy*, 18(12), 1013–1018.

Diéguez, P. M., Urroz, J. C., Marcelino-Sádaba, S., Pérez-Ezcurdia, A. et al. (2014), Experimental study of the performance and emission characteristics of an adapted commercial four-cylinder spark-ignition engine running on hydrogen–methane mixtures, *Applied Energy*, 113, 1068–1076.

Ganesh, R. H., Subramanian, V., Balasubramanian, V., Mallikarjuna, J. M. et al. (2008), Hydrogen fueled spark-ignition engine with electronically controlled manifold injection: An experimental study, *Renewable Energy*, 33(6), 1324–1333.

Gu, X., Huang, Z., Cai, J., Gong, J. et al. (2012), Emission characteristics of a spark-ignition engine fuelled with gasoline-n-butanol blends in combination with EGR, *Fuel*, 93, 611–617.

Heywood, J. B. (1988), Internal Combustion Engines Fundamentals, New York: McGraw-Hill, Inc.

Hinton, N. and Stone, R. (2014), Laminar burning velocity measurements of methane and carbon dioxide mixtures (biogas) over wide ranging temperatures and pressures, *Fuel*, 116, 743–750.

Huynh, T. C., Kang, J. K., Noh, K. C., Lee, J. T. et al. (2008), Controlling backfire for a hydrogen-fueled engine using external mixture injection, *Journal of Engineering for Gas Turbines and Power*, 130, 1–8.

Ibrahim, A. and Bari, S. (2008), Optimization of a natural gas SI engine employing EGR strategy using a two-zone combustion model, *Fuel*, 87, 1824–1834.

Liu, S., Cuty Clemente, E. R., Hu, T. and Wei, Y. (2007), Study of spark-ignition engine fueled with methanol/gasoline fuel blends, *Applied Thermal Engineering*, 27(11–12), 1904–1910.

Liu, X., Liu, F., Zhou, L, Sun, B. et al. (2008), Backfire prediction in a manifold injection hydrogen internal combustion engine, *International Journal of Hydrogen Energy*, 33, 3847–3855.

Mani Sarathy, S., Oßwald, P., Hansen, N. et al. (2014), Alcohol combustion chemistry, *Progress in Energy and Combustion Science*, 44, 40–102.

Mathur, H. B. and Das, L. M. (1991), Performance characteristics of a hydrogen fuelled S.I. engine using timed manifold injection, *International Journal of Hydrogen Energy*, 16(2), 115–127.

Mathur, H. B., Das, L. M. and Patro, T. N. (1992), Effects of charge diluents on the emission characteristics of a hydrogen fuelled diesel engine, *International Journal of Hydrogen Energy*, 17(8), 635–642.

Mathur, H. B. and Khajuria, P. K. (1984), Performance and emission characteristics of hydrogen fueled spark-ignition engine, *International Journal of Hydrogen Energy*, 9(8), 729–735.

MNRE (2014), Comparative evaluations of performance and mass emissions of an automotive passenger vehicles fueled with the enriched biogas using field trial tests. Project carried out by IIT Delhi by MNRE sponsorship.

Mohammadi, A., Shioji, M., Nakai, Y. et al. (2007), Performance and combustion characteristics of a direct injection SI hydrogen engine, *International Journal of Hydrogen Energy*, 32, 296–304.

Nagalingam, B., Duebel, F. and Schmillen, K. (1983), Performance study using natural gas, hydrogen supplemented natural gas and hydrogen in AVL research engine, *International Journal of Hydrogen Energy*, 8(9), 715–720.

Negurescu, N., Pana, C., Popa, M. G. and Cernat, A. (2011), Performance comparison between hydrogen and gasoline fuelled spark-ignition engine, *Thermal Science*, 15(4), 1155–1164.

Rice, R. W., Sanyal, A. K., Elrod A. C. and Bata, R. M. (1991), Exhaust gas emissions of butanol, ethanol, and methanol-gasoline blends, *Transactions of the ASME—Journal of Engineering for Gas Turbines and Power*, 113(3), 377–381.

Salvi, B. L. and Subramanian K. A. (2015), Experimental investigation and phenomenological model development of flame kernel growth rate in a gasoline fuelled spark-ignition engine, *Applied Energy*, 139, 93–103.

Salvi, B. L. and Subramanian, K. A. (2016a), Experimental investigation on effects of compression ratio and exhaust gas recirculation on backfire, performance and emission characteristics in a hydrogen fuelled spark-ignition engine, *International Journal of Hydrogen Energy*, 41(13), 5842–5855.

Salvi, B. L. and Subramanian K. A. (2016b), Experimental investigation on effects of exhaust gas recirculation on flame kernel growth rate in a hydrogen fuelled spark-ignition engine, *Applied Thermal Engineering*, 107, 48–54.

Sierens, R. and Verhelst, S. (2003), Influence of the injection parameters on the efficiency and power output of a hydrogen fueled engine, *Transactions of the ASME—Journal of Engineering for Gas Turbines and Power*, 195(3), 444–449.

Subramanian., V., Mallikarjuna, J. M. and Ramesh, A. (2007a), Effect of water injection and spark timing on the nitric oxide emission and combustion parameters of a hydrogen fuelled spark-ignition engine, *International Journal of Hydrogen Energy*, 32, 1159–1173.

Subramanian, V., Mallikarjuna, J. M. and Ramesh, A. (2007b), Intake charge dilution effects on control of nitric oxide emission in a hydrogen fueled SI engine, *International Journal of Hydrogen Energy*, 32, 2043–2056.

Subramanian, K. A., Mathad Vinaya, C., Vijay, V. K. and Subbarao, P. M. V. (2013), Comparative evaluation of emission and fuel economy of an automotive spark-ignition vehicle fuelled with methane enriched biogas and CNG using chassis dynamometer, *Applied Energy*, 105, 17–29.

Subramanian, K. A., Shukla, S., Athwe, M., Babu, M. K. G. et al. (2009), Study of flame characteristics of a spark-ignition engine for gasoline fuel, SAE International, Paper No. 2009-28-0028.

Szwaja, S. and Naber, J. D. (2010), Combustion of n-butanol in a spark-ignition IC engine, *Fuel*, 89(7), 1573–1582.

Turner, D., Xu, H., Cracknell, R. F., Vinod, N. et al. (2011), Combustion performance of bio-ethanol at various blend ratios in a gasoline direct injection engine, *Fuel*, 90(5), 1999–2006.

Varde, K. S. and Frame, G. M. (1984), A study of combustion and engine performance using electronic hydrogen fuel injection, *International Journal of Hydrogen Energy*, 9(4), 327–332.

Verhelst, S., Demuynck, J., Sierens, R. et al. (2010), Impact of variable valve timing on power, emissions and backfire of a bi-fuel hydrogen/gasoline engine, *International Journal of Hydrogen Energy*, 35, 4399–4408.

Verhelst, S. and Wallner, T. (2009), Hydrogen-fueled internal combustion engines, *Progress in Energy and Combustion Science*, 35, 490–527.

Wei, H., Zhu, T., Shu, G., Tan, L. et al. (2012), Gasoline engine exhaust gas recirculation—A review, *Applied Energy*, 99, 534–544.

White, C. M., Steeper, R. R. and Lutz, A. E. (2006), The hydrogen-fueled internal combustion engine: A technical review, *International Journal of Hydrogen Energy*, 31, 1292–1305.

Woolley, R. L. and Heiqriksen, D. L. (1977), Water induction in hydrogen-powered IC engines, *International Journal of Hydrogen Energy*, 1, 401–412.

7

Utilization of Biofuels in Compression-Ignition Engines

CI engines operate with dual cycle (thermodynamics cycle). Even though SI engines provide higher specific power output (possible with stoichiometric air-fuel ratio) than compression-ignition engines, the thermal efficiency of CI engines is higher than with SI engines. The main advantages of SI engines are higher power output and speed, and compact engine design. However, the engine's compression ratio is limited due to knock. The air-standard efficiency shows that the thermal efficiency of the engine increases with an increasing compression ratio. However, compression ratio and thermal efficiency are limited in spark-ignition engines. Therefore, compression-ignition engines get more momentum as they do not have the limitation of an increase in compression ratio. The compression-ignition engine operates generally at a compression ratio from 16:1 to 22:1, whereas spark-ignition engines are in the range of 4:1 to 11:1 depending on the octane number of fuels. Therefore, the thermal efficiency of compression-ignition engines is generally higher than that of spark-ignition engines.

7.1 General Working Principle of Compression-Ignition Engines

7.1.1 Suction Stroke

Only air is inducted through the intake manifold during a suction stroke and the engine does not have an air throttling system; therefore, the volumetric efficiency of the engine is higher than in spark-ignition engines. A long intake pipe called the intake manifold is attached to the engine cylinder head. Upstream of the manifold is connected with air filter and then exposed it to atmosphere. The downstream of the manifold is connected with engine cylinder head in which intake valve is located to allow and stop the airflow from the manifold. The airflow rate will vary depending on the valve lift. The valve is lifted up and down by a rocker arm activated by push rod. The intake airflow is maximum at maximum lift position, and vice versa. If the engine speed increases, the air flow rate increases, but will decrease beyond a certain engine speed, because the valve is a mechanical device with its own limitations in responding to open or close, as it needs a minimum threshold time to function. The air enters into the engine cylinder through a valve due to creating vacuum pressure inside the cylinder while moving the piston downward toward BDC. The geometry of a manifold is similar to a helical type, and a swirl plate is placed on the valve head toward the valve stem to create swirl motion of air. Swirl is an organized rotation of air about the cylinder axis. Swirl enhances the mixing rate of air with the injected fuel. The volumetric efficiency of any type of engine would be less than 100% mainly due to a decrease in the density of air by hot residual gas, effect of intake manifold geometry (including valve) on charge/air velocity, and blow-by losses. The in-cylinder temperature and pressure at BDC during a suction stroke is almost equal to the atmospheric pressure and temperature.

7.1.2 Compression Stroke

The inducted air is compressed by moving a piston toward TDC. The in-cylinder pressure and temperature of air increase during a compression stroke. The crank angle can be from 10° to 20° before TDC, and a high cetane number fuel is injected into the engine's cylinder using an injection system. The pump in the fuel tank supplies fuel to the injection pump, and the injection pump supplies the appropriate quantity of fuel at the desired injection pressure. Fuel with high injection pressure is directly injected into the cylinder by the injectors at the end of a compression stroke.

7.1.3 Injection Characteristics

The start of injection at the fuel pump is called static injection timing. The injection characteristics of a diesel engine include injection delay, static injection timing, dynamic injection timing (DIT) and injection duration. Injection delay is defined as the duration between dynamic injection timing and static injection timing. Dynamic injection timing is defined as the actual timing when the fuel is starting to inject into the cylinder, whereas static injection timing is always constant. Accurate control of operating fuel injection parameters (injection timing, injection delay, fuel rate shaping, and in-line fuel injection pressure) is necessary as it influences the mixture formation process of air and fuel. Fuel injection pressure is approximately in the range of 400 to 2200 bar. The fuel at the rate to be injected with respect to crank angle is called the rate shaping. The fuel rate is not same over the entire injection duration. In a normal mechanical injection pump, fuel flow rate is in an inverted U or V shape. Due to advancement of injection system technologies such as CRDI, the fuel rate can be designed as per the desired pattern. Fuel rate shaping influences fuel accumulation during the ignition delay period. Pulse injection or split injection is the total quantity of fuel divided into N pulses (N = number of pulses), and summation of all pulse quantity is the total fuel quantity to be injected per cycle. For example, if injectors in the injection system of a diesel engine is designed for three numbers of pulse injection, the total quantity is divided into three discrete quantities that may or may not be identical to each other. The first discrete quantity of fuel (first pulse) is injected from 10° crank angle (CA) before TDC (BTDC) to 8° CA BTDC, and the second and third pulses are from 6 to 1 CA BTDC and 1 to 4 CA after TDC (ATDC), respectively, and it may be noted that no fuel is injected from 8 CA BTDC to 6 CA BTDC or 1 CA BTDC to 1 CA ATDC. The combustion pattern will change with respect to this type of fuel injection rate shaping and this will enable better control of emissions (NOx and PM).

7.1.4 Fuel Spray Characteristics

The injection system of diesel engines influences spray characteristics (breakup length, spray cone angle, SMD, spray penetration, air entrainment, and vaporization) and the mixture formation process. Brake-up length is a fuel spray distance which is between from nozzle hole tip to a point where fuel diverging starts. Spray penetration is a measure of the depth of the fuel spray that is penetrated into the reactant (air) in the bowl of the combustion chamber during a compression stroke. SMD is the diameter of a fuel droplet that has the same surface-to-volume ratio as that of the total spray. Air entrainment is the process of air drawn into the fuel spray. The desired spray characteristics are less breakup length, optimum penetration distance, higher spray cone angle, less SMD, higher air entrainment, and higher vaporization rate of injected fuel droplets. These spray characteristics are influenced by injection characteristics and physico-chemical properties. The fuel spray characteristics influence the mixing characteristics of the injected fuel with the surrounding hot compressed air. If fuel spray characteristics are poor, the injected fuel does not mix properly with the surrounding air, which may result in a higher level of emissions and less thermal efficiency in a compression-ignition engine.

7.1.5 Ignition

Ignition initiates the chemical reaction of reactants (air-fuel mixture) and is an endothermic chemical process in a compression-ignition engine. In diesel engines, fuel at high pressure is injected into the engine cylinder using an injection system during the end of the compression stroke. The injected fuel has to be mixed with the surrounding air. Then, fuel droplets vaporize and subsequently a flammable mixture forms. The duration between the crank angles from the start of injection until the flammable mixture formation is called the physical ignition delay. The chemical reaction proceeds with the flammable mixture, and the duration between crank angles corresponding from the flammable mixture to the start of combustion is called chemical ignition delay. The time elapsed between the start of injection and combustion is called ignition delay. A diesel engine needs a minimum amount of time for the mixture preparation process. The cetane number of fuel, which indicates ignition quality of fuel, should be higher than 45 for initiation of ignition at a shorter time. The cetane number is related to self-ignition temperature and activation energy: if the cetane number is high, the self-ignition temperature is low, which means the fuel gets ignited at a lower temperature. On the other hand, if the cetane number is low, the self-ignition temperature is high, which results in longer ignition delay or the lower cetane number fuel needing a higher temperature to initiate ignition. Activation energy requirement of a fuel (gasoline, isooctane, isobutane, etc.) those molecule structure with olefin (unsaturated), aromatic (ring structure), and double bond is generally higher than a fuel with paraffin structure to break the bond to initiate a chemical reaction with an oxidizer compared to that of a paraffin-structured fuel (cetane, diesel, etc.), which is a straight chain that can easily break down into smaller molecules with relatively lower temperatures. Therefore, in general, iso-structured fuel needs more activation energy whereas paraffin-structured long-chain fuel (cetane, diesel, etc.) needs relatively less activation energy. However, there are a few fuels that are the exemption, since short molecular length paraffin fuel (e.g., methane) needs high activation energy to initiate ignition. The cetane number is correlated with the activation energy requirement of fuel. Heywood (1988) reported that the cetane number is inversely proportional to activation energy, as given in

$$CN = \frac{618840}{E_a} - 25 \tag{7.1}$$

Combustion proceeds with the accumulated fuel during the ignition delay period. The diffusion combustion phase is when combustion proceeds further with the injected fuel that is mixed with the remaining amount of oxygen along with partially and fully burned products. Heat is released and is converted to work by the movement of a piston toward BDC.

7.2 Emissions

Compression-ignition engines mainly emit the following emissions and are discussed individually.

7.2.1 Carbon Monoxide

Carbon monoxide forms as an intermediate product due to incomplete combustion. It is generally considered to have a two-step global reaction: the carbon in hydrocarbon fuel converts to carbon monoxide during combustion and then is further oxidized into carbon dioxide. However, in an actual chemical reaction, any number of local reactions could proceed to form intermediate products and final products.

7.2.2 Air-Fuel Ratio

If the air-fuel ratio is less than the stoichiometric air fuel ratio, which is called a rich mixture, the oxygen (in air) requirement for complete combustion is less than the theoretically required amount. For example, one mole of cetane ($C_{16}H_{34}$) needs 25 moles of oxygen for complete combustion, as shown in Equation 7.2. If 20 moles of oxygen are only available for combustion, carbon monoxide is one of intermediate products produced during combustion, as shown in Equation 7.3. Compression-ignition engines generally operate at a lean mixture and emit less CO emission than spark-ignition engines.

$$C_{16}H_{34} + 25(O_2 + 3.76N_2) \rightarrow 8CO_2 + 17H_2O + 94N_2 \tag{7.2}$$

$$C_{16}H_{34} + 20(O_2 + 3.76N_2) \rightarrow 9CO + 7CO_2 + 17H_2O + 94N_2 \tag{7.3}$$

CO emission is mainly a function of the air-fuel ratio (a/f), temperature (T), dissociation (D), poor mixture formation, and distribution in the combustion chamber (M):

$$CO = f\left(\frac{a}{f}, T, D, M\right) \tag{7.4}$$

7.2.3 Lower In-Cylinder Temperature

Even if the air-fuel ratio is stoichiometric or a lean mixture, CO emission would form in internal combustion engines. The reason for the CO formation is that if the in-cylinder temperature is lower due to poor combustion or more heat loss to the walls, the intermediate product of CO emission is unable to oxidize into CO_2. In general, during startup of engines, transient emissions including CO will be relatively higher mainly due to a lower cylinder temperature. CO emission will be relatively higher at idle, part load, and during the warm-up period compared to engine operation at higher loads.

7.2.4 Lower Degree of Homogeneous Charge

If injected fuel in the surrounding air is not mixed properly, CO will form wherever a rich mixture is present in the combustion chamber.

7.2.5 Dissociation of CO_2 at High Temperature

Even if an engine operates with the correct air-fuel ratio, perfect air-fuel mixing, and high temperature, CO can form due to dissociation of some of the CO_2 into CO, as shown in

$$xCO_2 \rightarrow aCO + bCO_2 \quad \text{(at high temperature)} \tag{7.5}$$

7.2.6 HC

CO and HC emissions are interlinked with each other as both emissions form mainly due to incomplete combustion. If an engine operates at a rich fuel-air mixture, it emits a high level of CO as well as HC emissions. As well, HC emission also forms due to hydrocarbon fuel trapped in a crevice volume (the gap between a piston and a cylinder liner), flame quenching due to temperature difference between flame/burning species and cylinder wall, wall impingement (fuel impingement) on piston and cylinder wall, SAC volume/area in injector hole, lubricating oil burning, and ultra-lean mixture (beyond flammability limit). HC emission is generally lower in compression-ignition engines mainly due to lean mixture operation.

7.2.7 NOx

Combined together, NO and NO_2 is called NOx emission. It forms when inducted nitrogen reacts with oxygen at high temperature during combustion, which is called thermal NOx. N_2 reacting with HCN radicals forms NOx emission called prompt NOx. Fuel-bound NOx is when fuel containing nitrogen reacts with oxygen during combustion. Thermal NOx is a major contributor to emission formation, and NOx is the main function of temperature, oxygen availability, diluents, and reaction time. If in-cylinder temperature, oxygen availability, reaction time, or any combination thereof is high, NOx emission will be high. If the oxygen concentration is diluted with inert gas (exhaust gas recirculation), NOx emission would be reduced significantly. In compression-ignition engines, NOx emission is high due to lean mixture, high in-cylinder temperature due to heterogeneous charge, and more reaction time due to less cycle time (the speed of a CI engine is relatively lower than an SI engine). The NOx is mainly a function of the air-fuel ratio (a/f), temperature (T), oxygen availability (O_2), and residence time (RR), as shown in

$$NOx = f\left(\frac{a}{f}, \; T, \; O_2, \; RR\right) \tag{7.6}$$

7.2.8 PM

PM is defined as the combustion-generated particles comprised of organic (Os) and inorganic substances (IOs), which are generally in a liquid or solid state or a combined binder state. Organic substances include soot, hydrocarbon, and intermediate products. Soot is defined as a carbonaceous particle impregnated with tar material. Inorganic substances include sulfur, ketone and metals (from lubricants). PM is described in

$$\text{PM} = \text{Organic Substance} + \text{Inorganic Substance} \tag{7.7}$$

PM forms in compression-ignition engines mainly due to engine operation with a rich mixture, a poor mixture distribution in the combustion chamber, lower in-cylinder temperature, and oxidation/melting of an inorganic substance.

The main biofuels for compression-ignition engines are detailed next.

7.3 Utilization of Biodiesel in Compression-Ignition Engines

The effect of biodiesel as a fuel on injection, spray, performance, and emissions characteristics of compression-ignition engines are discussed below.

Biodiesel has a higher cetane number than that of base diesel and this fuel is more suitable for compression-ignition engines.

7.3.1 Effect of Biodiesel-Diesel Blends on Injection and Spray Characteristics of a Diesel Engine

Injection, spray, mixing, ignition, and combustion characteristics of a compression-ignition engine influence the engine's performance and emissions characteristics. Even though the name "biodiesel" implies similar properties of petroleum diesel, biodiesel's physicochemical properties (density, viscosity, distillation properties, bulk modulus, surface tension, etc.) are not exactly same as those of diesel; hence, the injection and spray characteristics of an engine with biodiesel will differ.

The fuel property known as the bulk modulus indicates the compressibility of the fuel. The start of injection timing of the diesel engine advances when the engine is fueled with biodiesel due to a bulk modulus (1500 MPa) that is higher than that of base diesel (1350 MPa) (Tat and Gerpen, 2003). However, Szybist et al. (2007) reported that higher fuel density of biodiesel advances the start of injection timing. The injection duration, injection timing, and injection pressure increased with increasing the percentage of biodiesel in diesel (B25, B50, B75, and B100) (Kegl and Hribernik, 2006). The higher bulk modulus of fuel would increase the in-line fuel injection pressure, resulting in higher spray penetration (Kegl and Hribernik, 2006).

Lapuerta et al. (2008b) reported that the reason for high injection pressure with biodiesel is due to reduced fuel leakage in the pump. The relation between bulk modulus with a change in in-line fuel pressure and volume is shown in Equations 7.8 and 7.9. Equation 7.10 shows that the injection pressure is directly proportional to bulk modulus of biodiesel.

$$B = k \times \left[\frac{dP}{\left(\dfrac{dV}{V} \right)} \right] \tag{7.8}$$

Equation 7.7 can be rearranged, as shown in Equation 7.8, to

$$dP = \frac{B}{k} \times \left(\frac{dV}{V} \right) \tag{7.9}$$

If the volumetric strain (dV/V) is constant, the pressure (dP) is directly proportional to bulk modulus (B), as shown in

$$dP \propto (B) \tag{7.10}$$

Higher fuel pressure with biodiesel-diesel blends may result in an increasing fuel-mass flow rate and spray penetration. This may lead to increased soot formation, as the core of a fuel spray is a one of the main sources of soot formation. Soot emission can be decreased by optimization of the diameter of an injector's hole (Dodge et al., 2002).

Breakup length is mainly a function of density and is higher with biodiesel than base diesel as the density of biodiesel is higher than base diesel. Breakup length can be decreased by increasing the injection pressure and decreasing the nozzle hole diameter (Hiroyasu, 1998; Shimizu et al., 1984). Agarwal and Chaudhury (2012), Hiroyasu (1998), Som et al. (2010), and Wang et al. (2010) reported that spray cone angle is less with biodiesel due to its higher viscosity. The SMD of an injected particle influences the mixing rate. If a droplet's size (SMD) is larger, the heat transfer rate from the surrounding air is less, resulting in poor vaporization. The SMD is higher with biodiesel due to its higher density, viscosity, and surface tension (Park et al., 2009; Som et al., 2010; Wang et al., 2010). The Weber number is also less due to the higher surface tension of biodiesel that indicates a larger SMD (Lee et al., 2005). The spray penetration is higher with all biodiesel-diesel blends than with base diesel. Higher spray penetration is beneficial for air entrainment but it may lead to wall impingement. However, higher too much spray penetration would lead to wall impingement on piston bowl and cylinder wall whereas under spray penetration would lead to poor air utilization in fuel spray (Lahane and Subramanian, 2014; Lahane and Subramanian, 2015).

Higher SMD and spray penetration with less spray cone angle occurs with biodiesel, resulting in poor atomization (Gao et al., 2009). The effect of fuel viscosity on SMD is higher than on fuel density (Ejim et al., 2007).

Fuel spray penetration with biodiesel is higher due to an increase in injection pressure (Kegl and Hribernik, 2006; Rajalingam and Farrell, 1999). Spray penetration can be decreased by optimizing the injector nozzle hole (Hiroyasu and Miao, 2003; Kuti et al., 2013; Ohrn et al., 1991). Fuel injection pressure, air density, and nozzle diameter influence spray penetration (Myong et al., 2004; Siebers et al., 1998). Optimization of injection and spray characteristics are necessary for better mixture preparation and hence less PM (Minato et al., 2005; Nishida et al., 2007; Tao and Bergstand, 2008; Zhang et al., 2008a, 2008b). Fuel penetration needs to be optimum because under- and overpenetration may result in poor air entrainment and wall impingement (Karimi, 2007). Fuel vaporization is an important process that influences flammable mixture preparation. Vaporization is directly proportional to SMD that can be analyzed using the D^2 law, which is mainly a function of SMD, specific heat of gas, and latent heat of vaporization (Rakopoulos et al., 2006).

7.3.2 Wall Impingement

Wall impingement occurs when the injected fuel impinges on or adheres to the surface area of a combustion chamber mainly on the inner side of the piston bowl. The air velocity is much less or nil near the surface area of the combustion chamber compared to the core of the chamber; hence, the impinged fuel droplets do not mix properly with the surrounding air. In the case of biodiesel, the fuel penetration distance will increase due to higher in-line fuel pressure caused by the higher bulk modulus of biodiesel. This is a durability issue, as it will affect the mixture preparation process and will also lead to deposits on the combustion chamber's inner surface of the piston bowl. The droplet wall impingement may be either dry impingement or wet impingement. Therefore, it affects engine durability, performance, and emissions (Moreira et al., 2010). The spray-wall interaction also depends on the distance between the nozzle and the impinged surface, and exchange of momentum and energy to the hot surface, droplet size, wall surface roughness, and so forth (Ko and Arai, 2002; Montanaro et al., 2012). The Weber number (We) is an important parameter in the wall impingement, as given in

$$We = \frac{\rho_l \times d \times v_n^2}{\sigma} \tag{7.11}$$

where ρ_l is the liquid density (kg/m^3), d is the droplet diameter (m). v_n is the normal impingement velocity (m/s) component, and σ is the surface tension (N/m).

The Weber number is a measure of the relation between the kinetic energy of the droplet and its surface energy. A very large Weber number means that the influence of the surface tension can be neglected in the droplet impingement process. The Weber number can be used for defining the probability of wall impingement (Wachters and Westerling, 1966). Whether the wall liquid wets the surface depends on the wall temperature, wall material, roughness, and other factors (Wachters and Westerling, 1966). The wall impingement model developed by Naber and Reitz (1988) (based on sticking, reflecting, and sliding) is used for multidimensional engine simulation. Wall impingement may result in a rich localized mixture and hence higher unburned hydrocarbon emission (Miers et al., 2005). The above discussion is based on droplet impingement on a hot surface plate. Subhash et al. reported that the probability of wall impingement with biodiesel is higher due to an increase in injection pressure (Lahane and Subramanian, 2012).

The wall impingement problem could be overcome by increasing swirl velocity, which can be generated by directing air movement tangentially into the cylinder (Jimenez-Espadafor et al., 2012). Air swirl is defined as the organized rotational motion about the cylinder axis. The swirl that is required for the enhancement of mixing of the injected fuel with air in diesel engines is an organized movement of the air with a particular direction of flow and it assists the breaking up of the fuel jet. The probability of wall impingement could be minimized by reducing spray penetration, which can be reduced by the enhanced swirl ratio (SR). If the swirl ratio increased, the spray penetration will be decreased (Subramanian and Lahane, 2013).

7.3.3 Ignition

Ignition delay is defined as the time lapse between the start of injection (dynamic injection timing) and start of combustion. Total ignition delay comprises physical ignition delay and chemical ignition delay. Pedersen and Qvale (1974) developed a theoretical model for the physical ignition delay period in a direct injection diesel engine. The model was derived based on the single droplet calculation. Fuel quality parameters such as density, viscosity, and air temperature, SMD, and droplet velocity are the main influencing factors for the physical delay period, whereas cetane number influences the chemical ignition delay period of diesel engines. Ignition delay of the engine decreases with increasing injection velocity and decreasing droplet diameter (Kang et al., 2001). The total ignition delay period is generally expressed by the Arrhenius equation as given in Equation 7.12. According to the ignition delay equation, the effective parameters are in-cylinder pressure and temperature. Ignition delay period decreases with increasing in-cylinder pressure.

$$\tau = A \times P^{-n} \times \exp\left(\frac{E_a}{R \times T}\right) \tag{7.12}$$

where τ is the ignition delay, A is a constant, P and T are the in-cylinder pressure and temperature, respectively, R is the universal gas constant, and E_a is the activation energy.

Activation energy can be calculated with the given input of the cetane number (*CN*) using (Heywood, 1988)

$$E_a = \frac{618840}{(CN + 25)} \qquad (7.13)$$

Activation energy decreases with an increasing cetane number and vice versa. Ignition delay could be decreased by an increasing compression ratio (Hardenberg and Hase, 1979). Biodiesel fuel has generally a higher cetane number than petroleum diesel, leading to lower ignition delay in a compression-ignition engine.

7.3.4 Combustion Characteristics of Biodiesel-Fueled Compression-Ignition Engines

The combustion characteristics of biodiesel-fueled diesel engines are as follows: the heat release rate with biodiesel is higher than with diesel mainly due to the oxygen content in biodiesel (11–15 wt.%), which also results in less emissions (Suh et al., 2008). The start of combustion advanced in a diesel engine with biodiesel due to an advance in dynamic fuel injection timing because the fuel has a higher bulk modulus (Qi et al., 2009). Kegl (2006) who conducted experimental tests on a diesel engine with biodiesel under varied retarded injection timing (from 23°, 21°, 20°, 19°, 18°, and 17°CA BTDC) reported that peak in-cylinder pressure decreases with retarded injection timing, resulting in NOx emission reduction, but combustion duration increased. However, heat release is higher with advanced injection timing due to an increase in the premixed combustion phase by a larger ignition delay (Gunabalan et al., 2010). As the calorific value of biodiesel (≈38 MJ/kg) is less than that of petroleum diesel (≈44 MJ/kg), more fuel needs to be injected in order to maintain the required energy rate and the same power output. Hence, fuel injection duration has to be increased, which may result in larger combustion duration (Kegl and Hribernik, 2006). Ignition delay, which is an important parameter for influencing knock, noise, and emission formation, is less, resulting in a lower peak rate of pressure rise and peak heat release (Nagaraju et al., 2008; Qi et al., 2009; Zhang and Gerpen, 1996).

7.3.5 Performance and Emission Characteristics of Diesel Engines Using Biodiesel

Engine power dropped with biodiesel due to the fuel's lower heating value (Xue et al., 2011). A biodiesel-fueled engine emits less carbon-based emission due to a lower carbon content in biodiesel than in diesel fuel (diesel: 87 and biodiesel: 77.2 wt.%). Graboski and McCormick (1998) concluded that emissions (CO, HC, PM, and PAH) decreased with biodiesel-diesel blends (B10, B20, B30, B50, and B100) with increase in brake-specific fuel consumption (BSFC) and NOx emission compared to base diesel. Rakopoulos et al. (2004) examined the comparative effect of oxygen in air and fuel on BSFC and concluded that the effect is greater with oxygen enrichment with inducted air.

Lapuerta et al. (2008a) reviewed the literature on the effect of biodiesel on BSFC, brake thermal efficiency (BTE), NOx, PM, THC, and CO of a biodiesel-fueled diesel engine and concluded that the majority of the literature reported that BSFC and NOx increased with biodiesel, whereas other emissions decreased. The question is frequently asked about what is the optimum biodiesel-diesel blend. However, the answer is inconclusive, as the optimum biodiesel-diesel blends reported by many researchers are B10 (Bajpai et al., 2009; Muralidharan et al., 2004), B15, and B20 (Mahanta et al., 2006).

Biodiesel properties will vary from one type of feedstock to another. Biodiesel is produced from different feedstock, so its physicochemical properties will vary with feedstock. Lin et al. (2009) studied the performance and emissions of a diesel engine fueled with different feedstock methyl esters, such as soybean oil methyl ester (SOME), peanut oil methyl ester (PNOME), corn oil methyl ester (COME), sunflower oil methyl ester (SFOME), rapeseed oil methyl ester (ROME), palm oil methyl ester (POME), palm kernel oil methyl ester (PKOME), and waste fried oil methyl ester (WFOME). They reported that performance and emissions characteristics of the engine were not the same with biodiesel derived from different feedstocks. For example, NOx emission increased from 466 ppm with base diesel to 566 and 492 ppm SOME and PKOME, respectively, and the emission is the same with SOME and KOME. Similarly, smoke emission decreased from 22% with base diesel to 9% with SOME and 6% KOME. But in general, an engine's performance and emissions are almost the same with all biodiesel. Table 7.1 shows the percentage changes in emissions of different engines fueled with biodiesel. It can be observed from the table that CO, HC, and smoke emissions are less with biodiesel than with petroleum diesel. However, NOx emission is higher with biodiesel.

NO emission is formed at high temperature in the engine. NOx emission can be reduced by retarding of the injection timing, water injection/emulsion, EGR, selective catalyst reduction, and lean-burn operation. NOx emission is higher with biodiesel, so the emission needs to be reduced using the above techniques. Optimization of injection timing has to be done to achieve better performance (especially when the start of combustion nearby is TDC); however, NOx may go up. Therefore, injection timing has to be adjusted based on the trade-off between NOx and thermal efficiency without increasing other emissions. EGR is also a promising technology for NOx reduction at the source level in diesel engines.

Properties such as higher viscosity, high boiling point, and unsaturated hydrocarbon of biodiesel may lead to carbon deposit in an injector (coking), piston, and combustion chamber when the engine is operated for a longer time period. The injection quantity decreased about 26.26% due to blockage of fuel flow by corrosion of the injector hole (Karamangil et al., 2012). Fuel injection can fail with bio-oil due to plastic deformation, clogging of an injector's passage, erosion, and cavitation (Galle et al., 2012). Injector coking is more prone in diesel engines with straight vegetable oil (Pehan et al., 2009; Ramadhas et al., 2005). Injector coking and carbon deposits on combustion chamber in diesel engines are reported by Wander et al. (2011) when the engine was operated after 1000 hours. The reason is due to biodiesel properties such as higher viscosity, lower volatility, and unsaturated hydrocarbon compounds. Similarly, Liaquat et al. (2013) also conducted an endurance test (250 hours) and found injector coking.

The important points are summarized below:

- DIT of the engine advances at all loads with biodiesel due to the higher bulk modulus of biodiesel. Inline fuel injection pressure and injection duration increased with biodiesel.
- Spray penetration is higher with biodiesel/biodiesel-diesel blends. The spray breakup length, SMD, and air entrainment increased, whereas spray cone angle and percentage of vaporization decreased with all biodiesel-diesel blends at all loads.
- The probability of wall impingement is higher with biodiesel due to an increase in penetration distance.

TABLE 7.1

Comparison of Emissions of Different Diesel Engines Fueled with Biodiesel

Serial Number	Engine Specification	Feedstock for Biodiesel	CO	HC	NOx	Smoke	References
1.	Lombardini, four-stroke, single-cylinder, naturally aspirated, direct-injection variable-speed engine. Maxium power 8.1 KW at 3000 rpm.	Soybean	Biodiesel = 0.8% Diesel = 1.8% % change = 55.56% reduction with biodiesel	Biodiesel = 16 ppm Diesel =29 ppm % change = 44.8% reduction with biodiesel	Biodiesel = 1050 ppm Diesel = 860 ppm % change = 29.31% increase with biodiesel	Biodiesel = 8% Diesel = 15% % change = 46.66% reduction with biodiesel	Ozener et al., 2014
2.	Lombardini, four-stroke, single-cylinder, naturally aspirated, direct injection variable speed engine. Maximum power 8.1 KW at 3000 rpm.	Apricot seed kernel	Biodiesel = 0.05% Diesel = 0.18% % change = 72.2% reduction with biodiesel	NA	Biodiesel = 580 ppm Diesel = 410 ppm % change = 29.31% increase with biodiesel	Biodiesel = 8% Diesel = 42% % change = 57.14% reduction with biodiesel	Gumus et al., 2010
3.	Ford, four-stroke, six-cylinder, turbocharged, intercooled, direct injection engine. Variable-speed engine, 136 kW at 2400 rpm.	Waste frying oil	Biodiesel = 700 ppm Diesel = 1100 ppm % change = 36.36% reduction with biodiesel	Biodiesel = 25 ppm Diesel = 40 ppm % change = 37.5% reduction with biodiesel	Biodiesel = 1440 ppm Diesel = 1325 ppm % change = 8.67% increase with biodiesel	NA	Sanil et al., 2015
4.	Single-cylinder, four strokes, air-cooled, constant-speed engine (1500 rpm).	Karanja	Biodiesel = 0.5 g/ kW-hr Diesel = 2.68 g/ kW-hr % change = 81.34% reduction with biodiesel	Biodiesel = 0.05 g/ kW-hr Diesel = 0.02 g/ kW-hr % change = 75% reduction with biodiesel	Biodiesel = 8.07 g/ kW-hr Diesel = 6.24 g/ kW-hr % change = 29.32% increase with biodiesel	Biodiesel = 15.4% Diesel = 51.9% % change = 70.32% reduction with biodiesel	Lahane et al., 2015

(Continued)

TABLE 7.1 (CONTINUED)

Comparison of Emissions of Different Diesel Engines Fueled with Biodiesel

Serial Number	Engine Specification	Feedstock for Biodiesel	CO	HC	NOx	Smoke	References
5.	Yanmar, four-stroke, single-cylinder, direct-injection, air-cooled, variable-speed engine.	Algae	Biodiesel = 1.6% Diesel = 2.8% % change = 42.86% reduction with biodiesel	NA	Biodiesel = 555 ppm Diesel = 412 ppm % change = 34.7% increase with biodiesel	NA	Saddam et al., 2017
6.	Isuzu 4 HFI, inline four-cylinder engine. Maximum power 88 kW at 3200 rpm.	Waste cooking oil	Biodiesel = 237 ppm Diesel = 315 ppm % change = 24.76% reduction with biodiesel	Biodiesel = 141 ppm Diesel = 183 ppm % change = 22.95% reduction with biodiesel	Biodiesel = 236 ppm Diesel = 199 ppm % change = 18.59% increase with biodiesel	Biodiesel = 18.1% Diesel = 34.9 mg/m^3 % change = 48.14% reduction with biodiesel	Man et al., 2016
7.	Ricardo EG, four-stroke, single-cylinder engine. Maximum speed 3000 rpm.	Waste fish oil	Biodiesel = 600 ppm Diesel = 900 ppm % change = 33.33% reduction with biodiesel	Biodiesel = 27.5% Diesel = 7% % change = 74.54% reduction with biodiesel	Biodiesel = 306 ppm Diesel = 268 ppm % change = 14.18% increase with biodiesel	NA	Gharehghani et al., 2017
8.	Kirloskar, four-stroke, single-cylinder, constant speed (1500 rpm), air-cooled engine. Maximum power 5.22 kW.	Jatropha biodiesel and turpentine oil.	Biodiesel = 2.8 g/kW-hr Diesel = 3.2 g/kW-hr % change = 12.5% reduction with biodiesel	Biodiesel = 0.28 g/kW-hr Diesel = 0.19 g/kW-hr % change = 32.14% reduction with biodiesel	Biodiesel = 6.4 g/kW-hr Diesel = 4.1 g/kW-hr % change = 56.09% increase with biodiesel	Biodiesel = 31.1% Diesel = 49.2% % change = 36.79% reduction with biodiesel	Dubey et al., 2017

(Continued)

TABLE 7.1 (CONTINUED)

Comparison of Emissions of Different Diesel Engines Fueled with Biodiesel

Serial Number	Engine Specification	Feedstock for Biodiesel	CO	HC	NOx	Smoke	References
9.	Kirloskar, four-stroke, single-cylinder, constant-speed (1500 rpm), air-cooled engine. Maximum power 5.22 kW.	Cotton	Biodiesel = 545 ppm Diesel = 635 ppm % change = 14.17% reduction with biodiesel	NA	Biodiesel = 85 ppm Diesel = 69 ppm % change = 15.94% increase with biodiesel	NA	Shehata et al., 2013
10.	Kirloskar, four-stroke, single-cylinder, constant-speed (1500 rpm), air-cooled engine. Maximum power 5.22 kW.	Palm	Biodiesel = 615 ppm Diesel = 635 ppm % change = 3.14% reduction with biodiesel	NA	Biodiesel = 80 ppm Diesel = 69 ppm % change = 57.97% increase with biodiesel	NA	Shehata et al., 2013
11.	Kirloskar, four-stroke, single-cylinder, constant-speed (1500 rpm), air-cooled engine. Maximum power 5.22 kW.	Flax	Biodiesel = 500 ppm Diesel = 635 ppm % change = 21.25% reduction with biodiesel	NA	Biodiesel = 109 ppm Diesel = 69 ppm % change = 55.56% reduction with biodiesel	NA	Shehata et al., 2013

- Ignition delay and rate of pressure rise (RPR) decreased, whereas combustion duration increased with biodiesel.
- CO, HC, and smoke emissions decreased drastically with B100.
- NOx emission increased with biodiesel-diesel blends/biodiesel (B100).

7.4 Utilization of F-T Diesel in Compression-Ignition Engines

Biodiesel is an oxygenated fuel and has a lower calorific value due to its oxygen content, resulting in higher NOx emission and power drop. In the case of F-T diesel fuel, this fuel has superior physicochemical properties, such as a higher cetane number, higher calorific value (no oxygen embedded with diesel), and less density than base diesel. This fuel has more paraffin compounds and can be produced through the F-T synthesis process. The synthesis gas, which consists mainly of carbon monoxide and hydrogen, is produced from a variety of biomass feedstocks through the gasification process. The carbon and hydrogen in the biomass are converted into carbon monoxide and hydrogen through thermal pyrolysis of biomass with limited oxygen/air. Then, the synthesis gas is provided as feedstock to the F-T synthesis process. The final product includes straight-chain hydrocarbon, methanol, and DME. The performance, combustion, and emissions characteristics of F-T–diesel-fueled compression-ignition engines are summarized and given in Table 7.2.

7.5 Utilization of DME in Compression-Ignition Engines

The following conclusions can be summarized from the literature information.

- Specific fuel consumption of a DME-fueled compression-ignition engine with an unmodified fuel injection system increased with power drop due to its lower heating value as brake power at 4200 rpm dropped about 32.1% and BSFC at 2200 rpm increased about 47.1% (Sezer, 2011). Therefore, the injection system needs to be modified for injecting a large amount of DME in order to maintain the same power output. Fuel consumption will increase due to the lower heating value of DME (Huang et al., 2009).
- The oxygen content in DME fuel improves the combustion process, resulting in lower emissions. The ignition delay of DME is shorter than that of diesel fuel due to the higher cetane number of DME.
- Combustion duration is less with DME than that of diesel fuel as combustion is improved due to better mixing of the fuel with air (gaseous fuel), high vapor pressure, low boiling temperature, and oxygen content.
- NOx and HC emissions are decreased significantly and smoke emission is almost zero at all loads. However, CO emission increased slightly compared to diesel fuel (Huang et al., 2009). Ying reported that in DME- (10%, 20%, and 30% by mass) blended, diesel-fueled compression engines, smoke emission decreases

TABLE 7.2

Effect of F-T Diesel on Spray, Performance, Combustion, and Emissions of Compression-Ignition Engines Compared to Base Diesel

Serial Number	Characteristics	Reasons
I	**Spray**	
ii	Penetration distance	Less with F-T diesel than with base diesel due to lower T90 of F-T, causing faster evaporation than diesel (Gill et al., 2011)
iii	Air entrainment	Less with F-T diesel due to lower T90 of F-T diesel (Gill et al., 2011)
iv	Vaporization	Higher due to lower distillation temperatures of F-T diesel fuels (Gill et al., 2011)
II	**Performance**	
i	Power and torque output	Decreases due to lower heating value on a volumetric basis compared to diesel (Gill et al., 2011)
ii	Brake thermal efficiency	Slightly higher due to the low specific fuel consumption (due to high CV on gravimetric basis) with F-T diesel (Gill et al., 2011; Tao et al., 2007)
III	**Combustion**	
i	Ignition delay	Less due to the higher cetane number of F-T diesel than that of diesel (Gill et al., 2011; Tao et al., 2007)
ii	In-cylinder pressure and temperature	Less due to the higher cetane number of F-T diesel and less ignition delay (Tao et al., 2007)
iii	Rate of pressure rise	Less due to the higher cetane number of F-T diesel (Gill et al., 2011; Tao et al., 2007)
iv	Heat release rate and cumulative heat release	Peak HRR is slightly lower than diesel due to reduced ignition delay (Gill et al., 2011; Tao et al., 2007)
v	Combustion duration	Higher due to reduced ignition delay (Gill et al., 2011; Tao et al., 2007)
IV	**Emissions**	
i	Carbon monoxide	Less due to lower T90 and lower C/H ratio, which improves the combustion process (Gill et al., 2011; Tao et al., 2007)
ii	Hydrocarbon	Less due to lower T90 and lower C/H ratio and better combustion (Gill et al., 2011; Tao et al., 2007)
iii	Oxides of nitrogen	Less due to lower in-cylinder temperature with F-T diesel (Gill et al., 2011; Tao et al., 2007)
iv	Carbon dioxide	Less due to lower C/H ratio (Gill et al., 2011)
v	Soot/smoke/particulate matter	Less due to lower or nil aromatic and sulfur content in F-T diesel (Gill et al., 2011; Tao et al., 2007)

Abbreviation:　HRR, heat release rate.

significantly while NOx emission decreases moderately and CO and HC emissions increased at most operating conditions (Ying et al., 2006).

- Ying et al. (2006) conducted experiments on a 2.0878-liter diesel engine fueled with DME (Wang et al., 2014). They reported that BSFC decreases with an increase in DME quantity in diesel. NOx and smoke emissions decrease with an increase in DME quantity. However, CO and HC emissions increase with an increase in a DME blend.

- DME (CH_3-O-CH_3) only has C-H and C-O bonds and no C-C bond with high oxygen content (34.8%) by weight, may result in less smoke/PM emission.

TABLE 7.3

Summary of DME Utilization in Diesel Engines

Serial Number	Characteristics	Reasons
i	Spray	
	Breakup length	Decreases due to lower boiling point, lower density, and higher volatility of DME compared to that of diesel (Arcoumanis et al., 2008; Hyun et al., 2008)
	SMD	Smaller due to the lower kinematic viscosity of DME, which enhances fuel atomization (Hyun et al., 2008)
	Penetration distance	Decreases due to lower boiling point, lower density, and higher volatility of DME than that of diesel (Arcoumanis et al., 2008; Hyun et al., 2008)
	Vaporization	Faster vaporization due to lower boiling point, lower density, and higher volatility of DME compared to diesel (Arcoumanis et al., 2008; Hyun et al., 2008)
ii	Combustion	
	Ignition delay	Decreases due to low boiling point and high volatility of DME (Arcoumanis et al., 2008)
	In-cylinder pressure and temperature	Decreases due to fast evaporation of DME (Crookes et al., 2007)
iii	Emissions	
	Carbon monoxide	Decreases due to the low C/H ratio, high oxygen content, and the absence of C–C bonds in DME that would provide faster and more effective oxidation (Arcoumanis et al., 2008); however, some researchers reported the opposite trend (Crookes et al., 2007)
	Hydrocarbon	Significantly reduces due to less fuel-rich regions caused by shorter ignition delay and presence of oxygen in the DME (Teng et al., 2001); however, some researchers reported the opposite trend (Crookes et al., 2007)
	Oxides of nitrogen	Decreases due to shorter ignition delay with DME (Arcoumanis et al., 2008)
	Carbon dioxide	No significant change (Crookes et al., 2007)
	Soot/smoke/particulate matter	Almost zero due to the presence of oxygen content and absence of C-C bonds in DME (Arcoumanis et al., 2008)

- DME has a low boiling point (–25°C or 248K) that vaporizes easily in a cylinder when injected in a liquid phase and leads to better atomization and improved combustion even during cold weather conditions.

- The low viscosity of DME compared with diesel causes leakage from storage, a fuel supply system that relies on small clearances for sealing, and causes an increased amount of leakage in fuel pumps and fuel injectors.

- DME has a lower calorific value that is 64.7% of that of diesel and lower density, thus requires a higher volumetric flow and higher injection period for DME to produce the same output as diesel.

- The handling of DME is easy and comparable to LPG since DME vapor pressure is 530 kPa at 298K and thus liquefies easily at relatively low pressure.

- DME has low lubricity and can cause intensified surface wear of moving parts with a fuel injection system. Therefore, DME fuel is not compatible with conventional fuel delivery and fuel injection systems.

- DME fuel is not compatible with certain elastomers used in conventional diesel fuel system due to its corrosiveness. DME can be chemically reactive with sealing materials, plastic components, and fuel injection system. Therefore, a careful selection of sealing materials such as polytetrafluoroethylene (PTFE) is required to prevent deterioration after prolonged exposure to DME for DME-filled storage vessels, supply lines, and so forth.

- DME has a wider flammability limit in combustion system; thus, it requires adoption of rigorous procedures for safe operation.

A summary of the effects of DME fuel on spray, performance, combustion, and emissions characteristics of a diesel engine is given in Table 7.3

Summary

Since the cetane number of biodiesel is higher than that of petroleum diesel, in general, the cold-start ability and transient emissions of a compression-ignition engine with biodiesel would be better thanthe engine fueled with petroleum diesel. Emissions, such as CO, HC, and smoke/PM, are less in biodiesel-fueled compression-ignition engines as compared to base diesel. Less PM emission with biodiesel is due mainly to the absence of sulfur, aromatic components, and fuel-embedded oxygen. However, NOx emission in an engine with biodiesel is higher than base diesel and power output of the engine would drop with biodiesel.

F-T diesel is better than biodiesel due to its higher cetane number and calorific value. However, since the density of fuel is less than in biodiesel, injection system optimization (calibration) is necessary. Emissions (CO, HC, NOx, PM) in compression-ignition engines with F-T diesel are less than with petroleum diesel. The main advantages of F-T diesel over biodiesel are less NOx emission and power drop due to the absence of oxygen in F-T diesel. Storage of F-T diesel is similar to that of petroleum diesel but biodiesel has storage issues because it has poor cold-flow properties.

A compression-ignition engine with DME could emit almost zero smoke emissions due to better mixing property (high diffusivity) with air. CO, HC, and NOx emissions of an engine with DME fuel would be less than that of petroleum diesel. However, the fuel injection system needs to be calibrated/optimized as the fuel is stored in liquid form (e.g., LPG) and inducted/injected to the engine in vapor form. Power drop is another major issue due to lower heat content of the fuel. If DME fuel is compared with biodiesel and F-T diesel, it is the best in terms of emissions. However, it needs new infrastructure for production, storage, transportation, and dispensing, and these are obstacles to enabling its use in in-use (on-road) vehicles and also immediately implementing in vehicle fleets.

Solved Numerical Problems

1. Calculate the injection delay and injection duration for both static injection timing and dynamic injection timing of a biodiesel-fueled compression-ignition engine. The fuel starts to flow at the injection pump at 20 BTDC and it injects into the cylinder at 5 BTDC. The end of the fuel injection is at 10 ATDC.

Input Data

Take the crank angle from 1 CA to 180 CA for a suction stroke, 181 CA to 360 CA for a compression stroke, 361 CA to 540 CA for an expansion stroke, and 541 CA to 720 CA for an exhaust stroke

Static injection timing = 20 BTDC = 340 CA

Dynamic injection timing = 5 BTDC = 355 CA

End of injection timing = 10 ATDC = 370 CA

Injection delay = ?

Injection duration = ?

Solution

Injection delay = Start of actual injection by injector into the engine cylinder −

Start of fuel flow from injection pump

= 355 CA − 340 CA = 15 CA

Injection duration (static) = End of injection timing − Static injection timing

= 370 CA − 340 CA = 30 CA

Injection duration (dynamic) = End of injection timing − Dynamic injection timing

= 370 CA − 355 CA = 15 CA

2. The mass flow rate of diesel in a single-cylinder compression-ignition engine is 1 g/s. The speed of the engine is 3600 rpm and the engine's injector has five holes. If the engine with a conventional injection system (without modification) is to be fueled with biodiesel, calculate the change in the mass flow rate of biodiesel per cycle compared to diesel. The density and calorific value of diesel and biodiesel are 840 kg/m³ and 43 MJ/kg with diesel fuel and 860 kg/m³ and 38 MJ/kg with biodiesel diesel. The thermal efficiency and brake power of the engine are the same for both fuels (diesel and F-T diesel).

Input Data

Mass flow rate of diesel (m_f) = 1 g/s

Density of the fuel = 880 kg/m³ (biodiesel) and 840 kg/m³ (diesel)

Speed of the engine (N) = 3600 rpm

Calorific value of fuels = 38 MJ/kg (biodiesel) and 43 MJ/kg (diesel)

Number of holes in injector = 5

Change in mass flow rate of biodiesel per cycle = ?

Solution

$$m_{fc} = m_f \times 2 \times 60/N \left(\frac{kg}{cycle} \right)$$

Energy flow rate of diesel fuel = Mass flow rate of diesel × calorific value

$$= 1 \times 43500/1000 = 43.5 \, kJ/s$$

Because brake power and efficiency of the engine with diesel and biodiesel are the same in this problem, the energy flow rate of diesel fuel = Energy flow rate of biodiesel = 43.5 kJ/s

Mass flow rate of biodiesel = Energy flow rate/calorific value = 43.5 × 1000/38000 = 1.14 g/s

$M_{diesel} = 1 \times 2 \times 60/3600 = 0.033 \, g/cycle$

$M_{biodiesel} = 1.14 \times 2 \times 60/3600 = 0.038 \, g/cycle$

Increase in mass flow rate of biodiesel = $M_{biodiesel} - M_{diesel}$

$$= 0.038 - 0.033 = 0.005 \, g/cycle$$

Percentage change in mass flow rate of biodiesel = 0.005 × 100/0.033 = 15.2%

Inference

The mass flow rate of biodiesel has to be increased about 15% in order to maintain the same power output with the assumption of the same thermal efficiency. The injection duration needs to be increased if the biodiesel fuel injection rate is maintained at the same rate as diesel. Otherwise, the injection rate has to be increased if the same injection duration of diesel is maintained for biodiesel. In all cases, the power drop of the engine depends on the capacity of the fuel injection system whether it supplies the required higher quantity of biodiesel fuel or not.

3. A compression-ignition engine needs to be operated with biodiesel with a conventional injection system using the data below. If the fuel injection system can supply 10% of excess fuel, calculate the power drop of the engine. Brake thermal efficiency of the engine for both fuels is the same at 35%.

Input Data

The mass flow rate of fuel at a rated power output (m_f) = 1 g/s (diesel) and 1.14 g/s (biodiesel)

Capacity of fuel injection system = 1.1 times of the rated capacity

Thermal efficiency of the engine for both fuels is same at 35%

Take calorific value of biodiesel (CV) = 38,000 kJ/kg

Power drop of the engine = ?

Percent change in power drop of the engine = ?

Solution

The maximum diesel fuel supply by the conventional injection system = diesel fuel supply at rated capacity × (1+ excess quantity/100) = 1 × 1.1 = 1.1 g/s

The required quantity of biodiesel fuel has to be injected = 1.14 g/s

The maximum biodiesel quantity that can be injected by the injection system = 1.14 − 1.1 = 0.04 g/s

Brake power = efficiency × m_f × CV

Brake power at required biodiesel flow rate = 0.35 × 1.14 × 38000/1000 = 15.16 kW

Brake power at the possible biodiesel flow rate due to limitation of conventional fuel injection system = 0.35 × 1.1 × 38000/1000 = 14.63 kW

The drop in power output = Brake power at possible flow rate −

Brake power at required biodiesel flow rate

$$= 14.63 − 15.16 = −0.53 \, kW$$

Percent change in power output = (14.63 − 15.16)/15.16 = −3.5%

A biodiesel-fueled engine with the above condition would operate with a 3.5% drop in brake power.

4. An automotive compression-ignition engine delivers a rated power output of 5.5 kW. The engine is fueled with diesel and biodiesel. The bore and stroke of the engine are 86 mm and 75 mm, respectively. The injector hole diameter is 0.19 mm and the nozzle opening pressure of the injector is 250 bar. The density and viscosity of diesel are 830 kg/m³ and 2.64 cS, respectively, and the biodiesel is 860 kg/m³ and 5.2 cS. Calculate change in (i) breakup length, (ii) penetration distance, (iii) spray cone angle, (iv) Sauter mean diameter, and (v) air entrainment of fuel in a compression-ignition engine when the engine is running at 1500 rpm and the torque produced is 17 Nm. The data for diesel and biodiesel is given in the following table:

Fuel Name	Cylinder Pressure (Bar)	Line Pressure (Bar)	Density of Charge (kg/m³)	Fuel Volume (mm³/cycle)	Injection Timing (Degrees before TDC)
Biodiesel	36.866	304	232.68	44.65	11
Diesel	38.171	258.82	204.88	39.52	11

Solution

(i) Breakup length (l_b)

$$l_b = 15.8 \times \left(\frac{\rho_l}{\rho_g} \right)^{0.5} \times d_n$$

where, ρ_l is the density of fuel in kg/m³, ρ_g is the density of the charge in kg/m³ (calculated from $\rho_g = \dfrac{RT}{p}$), and d_n is the diameter of the nozzle

For diesel

$$l_b = 15.8 * \left(\frac{830}{204.88} \right)^{0.5} * .00019$$

$$l_b = .006 \, m$$

For biodiesel

$$l_b = 15.8 * \left(\frac{860}{232.68} \right)^{0.5} * .00019$$

$$l_b = .0057 m$$

(ii) Penetration distance (S)

$$S = 0.39 \left(\frac{2\Delta P}{\rho_l} \right)^{0.5} t \tag{7.14}$$

$$S = 2.95 \left(\frac{\Delta P}{\rho_g} \right)^{0.25} (d_n t)^{0.5} \tag{7.15}$$

where

$$t_b = \frac{29\rho_l}{(\rho_g \Delta p)^{0.5}} d_n$$

ρ_l is the density of fuel in kg/m^3, ρ_g is the density of the charge in kg/m^3, t is time after start of the injection in s (taken as 1), dp is the discharge pressure (difference between in-line pressure and in-cylinder pressure), and d_n is the diameter of the nozzle in m.

If $t_b < t$, use Equation 7.14 otherwise Equation 7.15

Here, t_b is less than t; therefore, using Equation 7.14

S for diesel

$$S = 0.39 \left(\frac{2 \times 220.649}{830} \right)^{0.5} \times 1$$

$$S = .00681 \text{ m}$$

Similarly, s for biodiesel $S = 0.007377$ m

(iii) Spray cone angle (θ)

$$\tan \frac{\theta}{2} = \frac{0.289}{A} 4\pi \left(\frac{\rho_g}{\rho_l} \right)^{0.5}$$

Let us take $A = 4.11$ for this engine, ρ_l is the density of the fuel in kg/m^3, ρ_g is the density of the charge in kg/m^3, and is the injection timing in s

For diesel

$$\tan \frac{\theta}{2} = \frac{0.289}{4.11} * 4\pi \left(\frac{207.4}{830} \right)^{0.5}$$

$$\theta = 24.65°$$

Similarly, for biodiesel $\theta = 23.8°$

(iv) Sauter mean diameter (D_{sm})

$$D_{sm} = 23.9(\Delta p)^{-0.135}\rho_a^{0.121}V_f^{0.131}$$

where, ρ_l is the density of the fuel in kg/m^3, ρ_g is the density of the charge in kg/m^3, V_f is the volume of fuel delivered per cycle per cylinder in $mm^3/cycle$, and dp is the discharge pressure (the difference between in-line pressure and in-cylinder pressure (MPa))

For diesel

$$D_{sm} = 23.9(22.0649)^{-0.135}(207.4)^{0.121} \, 39.52^{0.131}$$
$$D_{sm} = 48.58 \text{ micron}$$

Similarly, for biodiesel 49.37 micron

(v) Air entrainment (m_a)

$$m_a = \frac{\pi}{3} \times \left(\tan\left(\frac{\theta}{2}\right) \right)^2 \times S^3 \times \rho_a$$

For diesel

$$m_a = \frac{\pi}{3} \times (0.415)^2 \times (0.00681)^3 \times 204.88$$
$$m_a = 1.183 \times 10^{-5} \text{kg} = 11.83 \text{ mg}$$

Similarly, for biodiesel, air entrainment is 35.319 mg.

5. A biodiesel-fuelled compression-ignition engine gets deposits inside of the inner wall of the injector hole. Twenty percent of the hole's area is filled with the carbon deposits (coking) and the fuel is injected through only 80% of the hole's area. Derive an equation for a reduction in the mass flow rate of fuel.

Solution

The mass flow rate of fuel is given by the following equation:

$$m_f = \rho A V$$

where, ρ is the density of the fuel in kg/m_3, A is the area of the nozzle hole, and V is the velocity of the fuel jet

If the area is reduced by 0.8, let the new area be $0.8\,A = A_1$

The diameter (d_1) for the reduced area A_1 can be calculated using the following equation:

$$d_1 = \sqrt{4 * \frac{.8A}{\pi}}$$

The spray velocity can be calculated using the following equation:

$$V = \sqrt{2 \times \frac{\Delta p}{\rho}}$$

As the area of the hole will be reduced, the change in pressure can be calculated using the following equation:

$$\Delta p = f \frac{l\rho v^2}{2d}$$

where $f = 64/Re$ and $Re = \dfrac{\rho VD^2}{\mu}$

where f is the fouling factor, ρ is the density of kg/m³, Re is Reynold's number, and μ is the viscosity of the fuel in centi stoke.

Therefore, Δp_1 can be written as:

$$\Delta p_1 = \frac{fl\rho v^2}{2 \times d_1}$$

Substituting d_1 in the above equation, the change in pressure for d_1 can be written as:

$$\Delta p_1 = \frac{fl\rho V^2}{2 \times \sqrt{\dfrac{3.2A}{\pi}}}$$

Thus, the velocity of the spray can be written as shown below:

$$V = \frac{\sqrt{2 \times \left(\dfrac{fl\rho V^2}{2 \times \sqrt{\dfrac{3.2A}{\pi}}} \right)}}{\rho}$$

The mass flow rate for the reduced area due to coking can be calculated using the following equation.

$$m_{f1} = \rho \times A \times \frac{\sqrt{2 \times \left(\dfrac{fl\rho V^2}{2 \times \sqrt{\dfrac{3.2A}{\pi}}} \right)}}{\rho}$$

The coefficient of discharge will also be changed but the change will be small and it can be neglected.

The reduction in mass flow rate $= m_f - m_{f1}$

$$m_f = \rho AV - \rho \times A \times \frac{\sqrt{2 \times \left(\dfrac{fl\rho V^2}{2 \times \sqrt{\dfrac{3.2A}{\pi}}} \right)}}{\rho}$$

$$m_f = \rho A \left\{ V - \frac{\sqrt{2 \times \left(\dfrac{fl\rho V^2}{2 \times \sqrt{\dfrac{3.2A}{\pi}}} \right)}}{\rho} \right\}$$

6. Based on numerical problem 4, the distance between the downstream of the diesel fuel spray and the surface of the piston bowl is in the range of 22 to 29 mm. The penetration distance for biodiesel is 7.38 mm. If the engine is operated with bio-diesel fuel, what is the probability of wall impingement?

Solution

The penetration distance for biodiesel is 7.38 mm, which is much lower than 29 mm, and therefore, there is no chance of impingement is nil.

7. Calculate the bulk modulus and change in the compressibility factor of biodiesel at injection pressure in a standard temperature and pressure (STP) condition, 450-bar pressure, and 1800 bar as compared to base diesel ($C_{16}H_{34}$). The density of diesel and biodiesel at STP is 810 and 880 kg/m³, respectively. Take: volume of the fuel (V) = and change in volume of the fuel (dv) = 2.56×10^{-5} m³ and -8.56×10^{-7} m³

Solution

$$\text{Bulk modulus} = -K \times \frac{dp}{\dfrac{dv}{v}}$$

Here K is the compressibility factor, dp is the discharge pressure (the difference between in-line pressure and in-cylinder pressure), and $\frac{dv}{v}$ is the volumetric strain in-cylinder.

K can be find out from the compressibility chart (please refer any standard chart/hand book of compressibility factor)

For 450 bar

Here, the Pr is less than 40:

$$Pr = pressure/critical\ pressure$$

For diesel and biodiesel according to the chart (please refer any standard chart/hand book of compressibility factor), the compressibility factor is 1.5 and 1.7, respectively.

V (instantaneous cylinder volume in m^3) and dv (rate of change in volume) as 2.56×10^{-5} m^3 and -8.56×10^{-7} m^3, respectively

$$\text{The bulk modulus for diesel} = -1.5 \times \frac{220.64}{\left[\dfrac{8.56 \times 10^{-7}}{2.56 \times 10^{-5}}\right]}$$

$$= 9897.86\ bar$$

Similarly, the bulk modulus for biodiesel = 11217.58 bar

References

Agarwal, A. K. and Chaudhury, V. H. (2012), Spray characteristics of biodiesel/blends in a high pressure constant volume spray chamber, *Experimental Thermal and Fluid Science*, 42, 212–218.

Arcoumanis, C., Bae, C., Crookes, R. and Kinoshita, E. (2008), The potential of di-methyl ether (DME) as an alternative fuel for compression-ignition engines: A review, *Fuel*, 87(7), 1014–1030.

Bajpai, S., Sahoo, P. K. and Das, L. M. (2009), Feasibility of blending karanja vegetable oil in petro-diesel and utilization in a direct injection diesel engine, *Fuel*, 88, 705–711.

Crookes, R. J., Bob-Manuel, K. D. H. (2007), RME or DME: A preferred alternative fuel option for future diesel engine operation, *Energy Conversion and Management*, 48(11), 2971–2977.

Dodge, L., Simescu, S., Neely, G. et al. (2002), Effect of small holes and high injection pressures on diesel engine combustion, SAE 2002 World Congress & Exhibition, Detroit, SAE No. 2002-01-0494.

Dubey, P. and Gupta, R. (2017), Effects of dual bio-fuel (Jatropha biodiesel and Turpentine oil) on a single cylinder naturally aspirated diesel engine without EGR, *Applied Thermal Engineering*, 115, 1137–1147.

Ejim, C. E., Fleck, B. A. and Amirfazli, A. (2007), Analytical study for atomization of biodiesels and their blends in a typical injector: surface tension and viscosity effects, *Fuel*, 86, 1534–1544.

Galle, J., Verhelst, S., Sierens, R. et al. (2012), Failure of fuel injectors in a medium speed diesel engine operating on bio-oil, *Biomass and Bioenergy*, 40, 27–35.

Gao, Y., Deng, J., Li, C. et al. (2009), Experimental study of the spray characteristics of biodiesel based on inedible oil, *Biotechnology Advances*, 27, 616–624.

Gharehghani, A., Mirsalim, M. and Hosseini, R. (2017), Effects of waste fish oil biodiesel on diesel engine combustion characteristics and emission, *Renewable Energy*, 101, 930–936.

Gill, S. S., Tsolakis, A., Dearn, K. D. et al. (2011), Combustion characteristics and emissions of Fischer–Tropsch diesel fuels in IC engines, *Progress in Energy and Combustion Science*, 37(4), 503–523.

Graboski, M. S. and McCormick, R. L. (1998), Combustion of fat and vegetable oil derived fuels in diesel engines, *Progress in Energy and Combustion Science*, 24(2), 125–164.

Gumus, M. and Kasifoglu, S. (2010), Performance and emission evaluation of a compression-ignition engine using a biodiesel (apricot seed kernel oil methyl ester) and its blends with diesel fuel, *Biomass and Bioenergy*, 34, 134–139.

Gunabalan, A., Tamilporai, P. and Ramaprabhu, R. (2010), Effects of injection timing and EGR on DI diesel engine performance and emission-using CFD, *Journal of Applied Sciences*, 10(22), 2823–2830.

Hardenberg, H. O. and Hase, F. W. (1979), An empirical formula for computing the pressure rise delay of a fuel from its cetane number and from the relevant parameters of direct-injection diesel engines, SAE No. 790493, 1823–1834.

Heywood, J. B. (1988), *Internal Combustion Engine Fundamentals*, McGraw-Hill International Editions, Automotive Technology Series, New York.

Hiroyasu, H. (1998), The structure of fuel sprays and the combustion processes in diesel engines, ASME ICE Fall Technical Conference, Paper No 98-ICE-117.

Hiroyasu, H. and Miao, H. (2003), Measurement and calculation of diesel spray penetration, ICLASS, 9th International Conference on Liquid Atomization and Spray Systems, http://www.ilasseurope.org/ICLASS/iclass2003/fullpapers/1413.pdf.

Huang, Z., Qiao, X., Zhang, W. et al. (2009), Dimethyl ether as alternative fuel for CI engine and vehicle, *Fronters of Energy and Power Engineering in China*, 3(1), 99–108.

Jimenez-Espadafor, F. J., Torres, M., Velez, J. A. et al. (2012), Experimental analysis of low temperature combustion mode with diesel and biodiesel fuels: A method for reducing NOx and soot emissions, *Fuel Processing Technology*, 103, 57–63.

Kang, S. H., Baek, S. W. and Choi, J. H. (2001), Auto-ignition of sprays in a cylindrical combustor, *Heat and Mass Transfer*, 44, 2413–2422.

Karamangil, M. I. and Taflan, R. A. (2012), Experimental investigation of the effect of corrosion in two piezo driven diesel injectors, *Journal of the Energy Institute*, 85(4), 209–219.

Karimi, K. (2007), Characterization of multiple-injection diesel sprays at elevated pressures and temperatures, PhD thesis, School of Engineering, University of Brighton, United Kingdom.

Kegl, B. (2006), Experimental investigation of optimal timing of the diesel engine injection pump using biodiesel fuel, *Energy & Fuels*, 20, 1460–1470.

Kegl, B. and Hribernik, A. (2006), Experimental analysis of injection characteristics using biodiesel fuel, *Energy & Fuels*, 20, 2239–2248.

Ko, K. and Arai, M. (2002), Diesel spray impinging on a flat wall, characteristics of adhered fuel film in an impingement diesel spray, *Atom Sprays*, 12, 737–751.

Kuti, O. A., Zhu, J., Nishida, K. et al. (2013), Characterization of spray and combustion processes of biodiesel fuel injected by diesel engine common rail system, *Fuel*, 104, 838–846.

Lahane, S. and Subramanian, K. A. (2012), Modelling and CFD simulation of effects of spray penetration on piston bowl impingement in a DI diesel engine for biodiesel-diesel blend (B20), Paper No. ICES2012-81171, Proceedings of the ASME 2012, ICES2012, 163–170.

Lahane, S. and Subramanian, K. A. (2014), Impact of nozzle holes configuration on fuel spray, wall impingement and NOx emission of a diesel engine for biodiesel–diesel blend (B20), *Applied Thermal Engineering*, 64, 307–314.

Lahane, S. and Subramanian, K. A. (2015), Effect of different percentages of biodiesel–diesel blends on injection, spray, combustion, performance, and emission characteristics of a diesel engine, *Fuel*, 139(1), 537–545.

Lapuerta, M., Armas, O. and Fernandez, J. R. (2008a), Effect of biodiesel fuels on diesel engine emissions, *Progress in Energy and Combustion Science*, 34, 198–223.

Lapuerta, M., Armas, O. and Fernandez, J. R. (2008b), Effect of the degree of unsaturation of biodiesel fuels on NOx and particulate emissions, *SAE International Journal of Fuels and Lubricants*, SAE No. 2008-01-1676, pp. 1150–1158.

Lee, C. S., Park, S. W. and Kwon, S. II. (2005), An experimental study on the atomization and combustion characteristics of biodiesel-blended fuels, *Energy and Fuels*, 19, 2201–2208.

Liaquat, A. M., Masjuki, H. H., Kalam, M. A. et al. (2013), Impact of palm biodiesel blend on injector deposit formation, *Applied Energy*, 111, 882–893.

Lin, B., Huang, J. and Huang, D. (2009), Experimental study of the effects of vegetable oil methyl ester on DI diesel engine performance characteristics and pollutant emissions, *Fuel*, 88, 1779–1785.

Mahanta, P., Mishra, S. C. and Kushwah, Y. S. (2006), An experimental study of *Pongamia pinnata* L. oil as a diesel substitute, *Proceedings of the Institution of Mechanical Engineers, Part A: Journal of Power Energy*, 20, 803–808.

Man, X. J., Cheung, C. S., Ning, Z. et al. (2016), Influence of engine load and speed on regulated and unregulated emissions of a diesel engine fuelled with diesel fuel blended with waste cooking oil biodiesel, *Fuel*, 180, 41–49.

Miers, S. A., Carl, L. A., Jason, R. B. et al. (2005), Impingement identification in a high speed diesel engine using piston surface temperature measurements, SAE World Congress & Exhibition, Detroit, Michigan, SAE No. 2005-01-1909.

Minato, A., Tanaka, T. and Nishimura, T. (2005), Investigation of premixed lean diesel combustion with ultra-high-pressure injection, SAE No. 2005-01-0914.

Montanaro, A., Malaguti, S. and Alfuso, S. (2012), Wall impingement process of a multi-hole GDI spray: Experimental and numerical investigation (No. 2012-01-1266). SAE Technical Paper.

Moreira, A. L. N., Moita, A. S. and Panao, M. R. (2010), Advances and challenges in explaining fuel spray impingement: How much of single droplet impact research is useful?, *Progress in Energy and Combustion Science*, 36, 554–580.

Muralidharan, M., Thariyan, M. P., Roy, S. et al. (2004), Use of pongamia biodiesel in CI engines for rural application, SAE No. 2004-28-0030, doi: 10.4271/2004-28-0030.

Myong, K., Arai, M., Suzuki, H. et al. (2004), Vaporization characteristics and liquid-phase penetration for multi-component fuels, SAE No. 2004-01-0529.

Naber, J. D. and Reitz, R. D. (1988), Modelling engine spray/wall impingement, SAE No. 880107, doi: 10.4271/880107.

Nagaraju, V., Henein, N., Quader, A. et al. (2008), Effect of biodiesel (B-20) on performance and emissions in a single cylinder HSDI diesel engine, SAE Technical Paper, SAE World Congress & Exhibition, SAE No. 2008-01-140.

Nishida, K., Zhang, W. and Manabe, T. (2007), Effects of micro-hole and ultra-high injection pressure on mixture properties of D.I. diesel spray, SAE No. 2007-01-1890, doi: 10.4271/2007-01-1890.

Ohrn, T. R., Senser, D. W. and Lefebvre, A. H. (1991), Geometrical effects on discharge coefficients for plain-orifice atomizers, *Atomization and Sprays*, 1(2), 137–153.

Ozener, O., Yuksek, L., Ergenç, A. T. et al. (2014), Effects of soybean biodiesel on a DI diesel engine performance, emission and combustion characteristics, *Fuel*, 115, 875–883.

Park, S. H., Kim, H. J., Suh, H. K. et al. (2009), A study on the fuel injection and atomization characteristics of soybean oil methyl ester (SME), *International Journal of Heat and Fluid Flow*, 30, 108–116.

Pedersen, P. S. and Qvale, B. (1974), A model for the physical part of the ignition delay in a diesel engine, SAE No. 740716, pp. 2625–2638, doi: 10.4271/740716.

Pehan, S., Jerman, M. S., Kegl, M. et al. (2009), Biodiesel influence on tribology characteristics of a diesel engine, *Fuel*, 88, 970–979.

Qi, D. H., Geng, L. M., Chen, H. et al. (2009), Combustion and performance evaluation of a diesel engine fueled with biodiesel produced from soybean crude oil, *Renewable Energy*, 34, 2606–2613.

Rajalingam, B. V. and Farrell, P. V. (1999), The effect of injection pressure on air entrainment in to transient diesel sprays, *SAE Journal of Engines*, 3, 652–660.

Rakopoulos, C. D., Antonopoulos, K. A., Rakopoulos, D. C. et al. (2006), Multi-zone modeling of diesel engine fuel spray development with vegetable oil, bio-diesel or diesel fuels, *Energy Conversion and Management*, 47, 1550–1573.

Rakopoulos, C. D., Hountalas, D. T., Zannis, T. C. et al. (2004), Operational and environmental evaluation of diesel engines burning oxygen-enriched intake air or oxygen-enriched fuels: A review, SAE No. 2004-01-2924.

Ramadhas, A. S., Jayaraj, S. and Muraleedharan, C. (2005), Characterization and effect of using rubber seed oil as fuel in the compression-ignition engines, *Renewable Energy*, 30, 795–803.

Saddam, H. A. and Talal, Y. (2017), Diesel engine performance and exhaust gas emissions using Microalgae *Chlorella protothecoides* biodiesel, *Renewable Energy*, 101, 690–701.

Sanli, H., Canakci, M., Alptekin, E. et al. (2015), Effects of waste frying oil based methyl and ethyl ester biodiesel fuels on the performance, combustion and emission characteristics of a DI diesel engine, *Fuel*, 159, 179–187.

Sezer, I. (2011), Thermodynamic, performance and emission investigation of a diesel engine running on dimethyl ether and diethyl ether, *International Journal of Thermal Sciences*, 50, 1594–1603.

Shehata, M. S. (2013), Emissions, performance and cylinder pressure of diesel engine fuelled by biodiesel fuel, *Fuel*, 112, 513–522.

Shimizu, M., Arai, M. and Hiroyasu, H. (1984), Measurements of breakup length in high speed jet, *Bulletin of the Japan Society of Mechanical Engineers*, 27(230), 1709–1715.

Siebers, D. L. (1998), Liquid-phase fuel penetration in diesel sprays, International Congress & Exposition, Detroit, Michigan, SAE No. 980809.

Som, S., Longman, D. E., Ramírez, A. I. et al. (2010), A comparison of injector flow and spray characteristics of biodiesel with petro diesel, *Fuel*, 89, 4014–4024.

Subramanian, K. A. and Lahane, S. (2013), Comparative evaluations of injection and spray characteristics of a diesel engine using karanja biodiesel-diesel blends, *International Journal of Energy Research*, 37, 582–597.

Suh, H. K. and Lee, C. S. (2008), Experimental and analytical study on the spray characteristics of dimethyl ether (DME) and diesel fuels within a common-rail injection system in a diesel engine, *Fuel*, 87(6), 925–932.

Suh, H. K., Roh, G. H. and Lee, C. S. (2008), Spray and combustion characteristics of biodiesel/diesel blended fuel in a direct injection common-rail diesel engine, *ASME Journal of Engineering for Gas Turbines Power*, 130, 2807–2815.

Szybist, J. P., Juhun, S., Mahabubul, A. and Boehman A. L. (2007), Biodiesel combustion, emissions and emission control, *Fuel Processing Technology*, 88, 679–691.

Tao, F. and Bergstand, P. (2008), Effect of ultra-high injection pressure on diesel ignition and flame under high-boost conditions, SAE No. 2008-01-1603.

Tat, M. E. and Gerpen, J. H. (2003), Measurement of biodiesel speed of sound and its impact on injection timing, Report by NREL/SR-510-31462.

Teng, H., McCandless, J. and Schneyer, J. (2001), "Thermochemical characteristics of dimethyl ether—An alternative fuel for compression-ignition engines," SAE Technical Paper 2001-01-0154.

Wachters, L. H. J. and Westerling, N. A. J. (1966), The heat transfer from a hot wall to impinging water drops in the spheroidal state, *Chemical Engineering Science*, 21(11), 1047–1056.

Wander, P. R., Altafini, C. R., Colombo, A. L. et al. (2011), Durability studies of mono- cylinder compression-ignition engines operating with diesel, soy and castor oil methyl esters, *Energy*, 36, 3917–3923.

Wang, X., Huang, Z., Kuti, O. A. et al. (2010), Experimental and analytical study on biodiesel and diesel spray characteristics under ultra-high injection pressure, *Heat and Fluid Flow*, 31, 659–666.

Wang, Y., Zhao, Y., Xiao, F. et al. (2014), Combustion and emission characteristics of a diesel engine with DME as port premixing fuel under different injection timing, *International Journal of Energy Conversion and Management*, 77, 52–60.

Wu, T., Huang, Z., Zhang, W. et al. (2007), Physical and chemical properties of GTL-diesel fuel blends and their effects on performance and emissions of a multicylinder DI compression-ignition engine, *Energy & Fuels*, 21, 1908–1914.

Xue, J., Grift, T. E. and Hansen, A. C. (2011), Effect of biodiesel on engine performances and emissions, *Renewable and Sustainable Energy Reviews*, 15, 1098–1116.

Ying, W., Longbao, Z. and Hewu, W. (2006), Diesel emission improvements by the use of oxygenated DME/diesel blend fuels, *Atmospheric Environment*, 40, 2313–2320.

Zhang, Y. and Gerpen, J. V. (1996), Combustion analysis of esters of soybean oil in a diesel engine, International Congress & Exposition, Detroit, Michigan, SAE No. 960765.

Zhang, W., Nishida, K. and Gao, J. (2008a), Experimental study on mixture formation process of flat wall impinging spray injected by micro-hole nozzle under ultra-high injection pressures, SAE International Powertrains, Fuels and Lubricants Congress, Shanghai, China, SAE No. 2008-01-1601.

Zhang, W., Nishida, K., Gao, J. et al. (2008b), An experimental study on flat-wall impinging spray of micro-hole nozzles under ultra-high injection pressures, *Proceedings of the Institution of Mechanical Engineers, Part D: Journal of Automobile Engineering*, 222, 1731–1741.

8

Biofueled Reactivity Controlled Compression-Ignition Engines

8.1 Introduction to Conventional Heterogeneous Charge, Premixed Charge, Homogeneous Charge, and Reactivity Charge

8.1.1 Heterogeneous Charge

Compression-ignition engines operate with a heterogeneous air-fuel mixture. A heterogeneous mixture means the air-fuel ratio or equivalence ratio in a spatial distribution of mixture is not the same in all parts of the combustion chamber. The diffusivity of high-cetane-number liquid fuels (diesel, biodiesel, F-T diesel, etc.) is less than that of gaseous fuels so these fuels have to be atomized using a high-injection, pressure-based injection system and subsequently the atomized fuel is mixed with the surrounding air, which is at a high temperature with intensified swirl and squish air motion. Then, the mixture starts to vaporize and is in an almost completely vapor state. A mixture prepared during the ignition delay period is called a premixed charge, whereas a mixture preparation at the end of the injection period is called a heterogeneous mixture. The available time duration for a premixed charge is relatively greater than in the later stage of preparation. The ignition delay of advanced engines is much less due to a better mixture formation with advanced technologies such as CRDI, even though the degree of homogeneous charge improves or the degree of heterogeneous charge decreases compared to conventional diesel injection technology. However, the degree of homogenous charge is the lowest with direct fuel injection based compression ignition engines compared to port injection or manifold injection based spark ignition engines because the time available to prepare mixture is lower with the direct fuel injection than manifold/port injection methods. For example, the mixture preparation time for a typical CI engine (injection duration from 10 CA to 20 CA at an engine speed of 3000 rpm) is in the range of 0.56 ms to 1.1 ms out of a cycle time of 40 ms. However, in the case of spark-ignition engines, the mixture preparation time will be nearly the half of a cycle time. The degree of homogenous charge can be increased by pulse injection or split injection because the fuel, which can be divided to any number of discrete quantities, is directly injected into the cylinder during the beginning of a compression stroke, and the time available for this type of injection is theoretically almost 180 CA. The injected fuel can mix properly due to more available time. However, control of autoignition (CAI) and combustion with knock with pulse injection are major technical challenges in a conventional compression-ignition engine. Note that pollutant formation in a CI engine is mainly due to the degree of charge homogeneity. PM forms less with a homogeneous mixture and more with a heterogeneous mixture. Even though advanced technology can drastically decrease the emission, it still will not meet the stringent emissions norms. It is a fact that the mass of PM is much less or negligible compared to the mass of fuel injected per cycle. The mass of PM is approximately less than 0.1% of fuel mass, but in terms of a particle's size and the number of counts of PM, it is high and unable to meet to present or future emissions norms. In addition, the localized temperature in a combustion chamber during combustion is higher with a heterogeneous mixture, resulting in a higher level of NOx. If the in-cylinder temperature is high, PM emission is lower but NOx emission will be high, and vice versa. Hence, a simultaneous reduction of NOx and PM is still challenge in compression-ignition engines. These emissions, along with specific fuel consumption, can be significantly reduced with a homogeneous charge. Because liquid fuels such as diesel, F-T diesel, and biodiesel are less volatile, an external mixture preparation in an engine's intake manifold during a suction stroke is not a desirable option. It is seen from the discussion that combustion with heterogeneous air-fuel in conventional diesel engines would lead to high NOx and smoke emissions. An LTC strategy with a premixed charge could reduce NOx and PM emissions because NOx formation needs high activation energy (Turns, 1999). In addition, soot emission with LTC will be reduced due to better mixing of a charge with a long ignition delay. However, the limitation with LTC is a less specific power output compared to conventional combustion in compression-ignition engines. Many researchers reported that NOx and smoke emissions in compression-ignition engines decreased simultaneously under HCCI and premixed charge compression ignition (PCCI) mode (Hardy et al., 2006; Opat et al., 2007).

Combustion with some types of homogenous charges is explained next.

8.1.2 PCCI

PCCI combustion is a hybrid combustion sandwiched between an HCCI and a conventional diesel combustion. The heat release rate is controlled by injection rate and diluents without knocking. Researchers (Noehre et al., 2006) from Lund University reported that a diesel engine under PCCI mode with 65% EGR at reduced compression ratio can reduce NOx and soot emissions simultaneously. Fuel wall wetting with early injection of diesel fuel is a technical issue as in-cylinder temperature, swirl, and squish are relatively less and diesel fuel may not be suitable for PCCI operation. Chen et al. (2014) reported a compression-ignition engine fueled with n-butanol-blended diesel with a medium EGR percentage could simultaneously reduce NOx and soot emissions. However, combustion with PCCI mode results in high levels of CO and unburned hydrocarbon (UHC) emissions. In addition, other problems associated with PCCI combustion, including low fuel volatility, high flammability of diesel, formation of a homogeneous mixture, CAI, limited operating range, and excessive wall impingement, need to be solved in order to obtain desirable benefits (Kokjohn et al., 2011; Yao et al., 2009).

8.1.3 HCCI

HCCI is a form of LTC in which a premixed charge gets autoignition in the engine cylinder. The HCCI concept was developed by combining the positive benefits of a homogeneous charge from spark-ignition engines and autoignition from compression-ignition engines. A compression ignition direction engine (CIDI) and an HCCI engine are shown in Figure 8.1. The main advantage of a spark-ignition engine is that combustion proceeds with a homogeneous charge, but a disadvantage is that flame combustion resulting in high localized in-cylinder temperature and hence NOx emission. Optimum flame velocity with respect to load and fuel quality is another challenge; otherwise, flame velocity is higher, resulting in

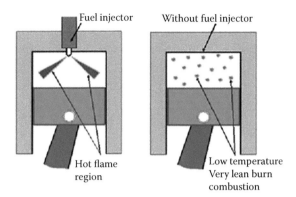

FIGURE 8.1
Representation of CIDI and HCCI.

flame quench and hence higher HC emissions and lubricating oil deterioration. On other hand, flame velocity is less, which may result in a high level of HC and CO emissions along with high specific fuel consumption. Therefore, in this regard, flame combustion is a disadvantage. In the case of a compression-ignition engine, combustion with autoignition is an advantage because all fuel molecules get combusted; however, combustion with a heterogeneous mixture is a major drawback in CI engines. Combined together, the salient aspects of a homogeneous charge (SI engine) and autoignition (CI engine) are known as HCCI. NOx and PM emissions along with specific fuel consumption in engines under HCCI mode will be less due to autoignition with a homogeneous charge.

Autoignition of a fuel-air mixture can be obtained by increasing the compression ratio (Aceves et al., 1999), variable valve actuation, preheating of induction gaseous fuels (Goran et al., 2004), and reinducted exhaust gases (Au et al., 2001). Saxena et al. (2013) reported that an engine under HCCI mode at high load will require turbochargers. Although HCCI combustion is thermodynamically attractive, CAI (ignition timing control and rate of heat release control) is a major challenge in terms of control of rapid heat release rate and rate of pressure rise.

8.1.4 HCCI Strategy in CI Engines

HCCI, which is a low-temperature combustion strategy, is generally used for simultaneous reduction of NOx and PM/smoke emissions in compression-ignition engines. For example, Helmantel and Denbratt (2004) reported that a diesel engine under HCCI mode could operate with a 98% and 95% reduction in NOx and smoke emissions, respectively, compared to conventional diesel mode. The degree of homogeneity can be increased by increasing a premixed charge that could result in simultaneous reduction of both NOx and PM emissions. NOx emission decreases with HCCI mode by reducing in-cylinder temperature due to enhanced charge homogeneity, whereas PM emission decreases due to better mixing of charge. Saxena and Bedoya (2013) reported that NOx and PM emissions decreased with LTC combustion. However, a higher premixed charge in the engine would lead to combustion with knocking. Bedoya et al. (2012) reported that a biogas-fueled CI engine (a rated power output of 60 kW at 3300 rpm) in HCCI mode could only operate with limited equivalence ratios of 0.25–0.4. Swami Nathan et al. (2010) reported that a 50% biogas energy share is the optimum share for better performance and lower emissions in HCCI mode in a 3.5-kW CI engine at 1500 rpm. Ibrahim and Ramesh (2013) reported that an engine (3.7 kW at 1500 rpm) fueled with hydrogen diesel in HCCI mode increased brake thermal efficiency from 30.2% in diesel-HCCI mode to 31.9% in hydrogen-diesel HCCI mode.

8.2 RCCI

RCCI, which is also called dual-fuel engine combustion technology, is a version of an HCCI-based compression-ignition engine. A compression-ignition engine in RCCI mode operates with two different kinds of fuels as high reactivity fuel for initiating ignition for combusting low reactivity fuel. High-cetane-number fuels, including diesel, biodiesel, F-T, and diesel are high reactivity fuels, whereas higher-octane-number fuels, including methanol, ethanol, natural gas, biogas, and gasoline are as low reactivity fuels. Higher-octane-number fuel, which generally has high volatility and can mix homogeneously with inducting air, is inducted using a low injection pressure system during a suction stroke. High reactivity fuel is injected at the end of a compression stroke using a conventional high injection pressure system, and the fuel first self-ignites (autoignition) due to a low self-ignition temperature and also acts as an ignition source for low reactivity fuels (high self-ignition temperature) that can't self-ignite.

The localized in-cylinder temperature is lower due to a homogenous charge, and dilution of oxygen concentration in the combustion chamber is greater with low reactivity fuel, resulting in less NOx formation. Better mixing of fuel with air leads to less formation of PM and possible higher thermal efficiency. The start of combustion in a CI engine in RCCI mode proceeds slowly hence rate of pressure rise and knocking are controlled. RCCI technology offers emission reduction at the source level as well as improved thermal efficiency.

The compression ratio of a high octane number fueled spark-ignition engine is limited due to knock and thermal efficiency. For example, the compression ratio of an SI engine for methane fuel is limited to 11:1. However, methane can be used in a CI engine in RCCI mode, which can operate at a compression ratio of more than 18:1, resulting in relatively higher thermal efficiency and less emissions compared to same fuel utilized in spark-ignition engines.

Mixing fuels with different autoignition properties can also decrease combustion speed (Hunter et al., 2005). Commercial gasoline (Reitz et al., 2015), methanol (Reitz et al., 2015) or ethanol (Reitz et al., 2015), or natural gas (Kalsi et al., 2016) can be used as low reactivity fuels in compression-ignition engines in RCCI mode. RCCI is a dual-fuel engine technology with at least two fuels of different reactivity. Li et al. (2014) reported simultaneous reduction of NOx and soot emissions in a compression-ignition engine in RCCI mode. The optimum methanol energy share with advanced start of injection was 66.5%. Dempsey et al. (2011) studied the use of high-octane gasoline in diesel engine in dual-fuel RCCI mode with optimum EGR percentages and swirl ratio, and they reported emissions reduction along with thermal efficiency improvement. Inagaki et al. (2006) conducted tests in RCCI mode with injection of isooctane into an intake port and diesel injected into an engine cylinder with advanced injection timing. The results showed that ignition timing can be controlled by optimizing the energy rate of two fuels. Professor Reitz from the Engine Research Center (ERC) at the University of Wisconsin Madison first refined RCCI engine technology through extensive experiment and simulation work (Splitter et al., 2010, 2011). Kokjohn et al. (2011) concluded that control of the ignition timing and HRR can be adjusted by adjusting the fuel reactivity of gasoline injected into the intake port and diesel is injected into the cylinder at advanced injection timing. They reported that a diesel-/gasoline-fueled RCCI engine can provide 53% indicated thermal efficiency and can meet the US-2010 regulations at low-to-medium load without any aftertreatment system. Sayin et al. (2009) carried out experimental tests on a methanol- (0% to 15%)/diesel-fueled dual-fuel diesel engine, and CO, HC, and smoke emissions from the engine decreased while NOx increased.

8.3 Case Study: Diesel/CNG and HCNG Fuel and Dual-Fuel RCCI Engine Mode

8.3.1 Dual-Fuel Engine

Gaseous fuels such as biogas, producer gas, natural gas, LPG, and hydrogen could be utilized in a CI engine in a dual-fuel mode. Banupurmath et al. (2008, 2009) studied the performance of a CI engine fueled with diesel, honge, rice bran, and neem oils with producer gas in a dual-fuel mode and they reported that smoke and NO emissions decreased but HC and CO emissions increased. HC and NOx emissions in a dual-fuel engine fueled with preheated refined rice bran oil (60°C), blended diesel (ratio of 3:1), and producer gas (53% by volume) decreased about 48% and 61%, respectively, but CO emission increased about 16.31% (Banupurmath et al., 2008, 2009). NOx emission in a compression-ignition engine (673cc swept volume) in a dual-fuel mode (natural gas-diesel) decreased significantly (Ryu, 2013). These gaseous fuels used in compression-ignition engines in a dual-fuel mode could reduce NOx and PM emissions. However, the disadvantages are lower brake thermal efficiency and higher HC and CO emissions (especially at part loads). These problems can be addressed using hydrogen in a dual-fuel engine. A brief review of the literature on hydrogen-based dual-fuel engines is given below.

Homan et al. (1979) from Cornell University studied utilization of 100% hydrogen in a CI engine with an increasing high compression ratio up to 29:1 and they reported that hydrogen cannot be ignited in the CI engine. Takahashi (1982) concluded that utilization of hydrogen in existing CI engines with compression ratios of 15.4:1 to 21:1 is difficult. The main reasons for the difficulty in using hydrogen in CI engines are high self-ignition temperature, higher octane number, and higher ignition delay as the temperature at which the energy available in hot air (by compression alone) is not sufficient to initiate ignition of the hydrogen-air charge. However, a small quantity of high-cetane-number fuel (diesel fuel, biodiesel, F-T diesel) could be used as an ignition source to initiate ignition and combustion of a premixed hydrogen-air mixture in a CI engine in a dual-fuel mode.

Wu and Wu (2013) concluded that thermal efficiency of a CI engine in a dual-fuel mode increased with a 30% hydrogen energy share at 100% load. Geo et al. (2008) also reported thermal efficiency improved from 29.9% with a base diesel mode to 31.6% with a 10.1% hydrogen energy share at 100% load.

At 100% load, HC and CO emissions decreased from 0.55 g/kWh and 3.14 g/kWh in a diesel mode to 0.44 g/kWh and 2.31 g/kWh with 10.1% hydrogen energy share in a dual-fuel mode (Geo et al., 2008). Tsolakis (2005) reported that total particle number and total particle mass decreased from 0.55 cm^{-3} and 98.66 mg/cm^3 with a base diesel mode to 0.37 cm^{-3} and 77.64 mg/cm^3 with 20% hydrogen with EGR in an 8.6-kW direct injection CI engine (8.6 kW, 1500 rpm) in a dual-fuel mode.

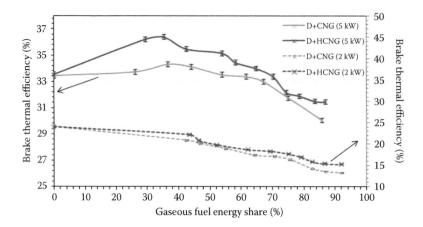

FIGURE 8.2
Brake thermal efficiency of a diesel engine in RCCI mode. (From Singh, S. (2017), Utilization of methane in a compression-ignition engine under dual fuel mode, PhD thesis, IIT Delhi.)

Hydrogen addition into a CI engine in a dual-fuel mode could lead to reduced GHG emissions (CO_2, methane (CH_4), and nitrous oxide (N_2O)) at the source level. CO_2 emission in a CI engine (11 kW rated power) decreased from 190 g/kWh in a base biodiesel mode to 104 g/kWh in a hydrogen-biodiesel dual-fuel mode (hydrogen consumption: 0.0005 kg/s) (Korakianitis et al., 2010). Methane emission, which is one of the GHG emissions, has a higher global warming potential of 25 (for a 100-year time horizon) than that of CO_2 emission. Subramanian and Chintala (2013) reported a methane emission reduction of about 22% with use of hydrogen (20% energy share) in a biodiesel-fueled CI engine. A CI engine (compression ratio of 19.5:1) could operate with a maximum of 19% hydrogen energy share in a dual-fuel mode for knock-free operation, and the maximum hydrogen energy share can be enhanced from 19% with a compression ratio of 19.5:1 to 59% and 63% with a reduction in compression ratios of 16.5:1 and 15.4:1, respectively (Chintala and Subramanian, 2015).

Experimental tests were conducted on a diesel engine (7.4 kW rated power output) in an RCCI mode (diesel-*). The higher-octane-number (low reactive) gaseous fuels (CNG and HCNG) were injected to the engine's intake manifold during a suction stroke, whereas diesel (high reactivity) pilot fuel was directly injected into the in-cylinder during the end of a compression stroke for initiating ignition and combustion. Experiments were conducted on a diesel engine in an RCCI mode at 5 and 2 kW electrical power outputs. The results indicate that thermal efficiency increased with HCNG (18% H_2 + 72% CNG) fuel (Figure 8.2) in an RCCI mode, and HC, CO, and smoke emissions decreased significantly (Figure 8.5). However, NOx emission in the engine with HCNG at 5-kW power output was higher compared to base diesel and base CNG (Figure 8.5). It can be observed from Figure 8.2 that the thermal efficiency of an engine with up to 60% CNG energy share is higher compared to base diesel mode. However, the efficiency at low power output (2 kW) is less with all gaseous fuels than with base diesel. The engine can operate up to about 85% CNG/HCNG energy share for 5 kW. Note that an engine can't operate with a 100% energy share in an RCCI mode as a minimum diesel energy share is needed to initiate ignition of the gaseous fuel charge.

---

* Compressed natural gas (CNG)/hydrogen blended compressed natural gas (HCNG).

Figure 8.3a, b, and c shows the variation of in-cylinder pressure, heat release rate, and in-cylinder temperature with respect to crank angle in the engine (5 and 2 kW) fueled with different CNG energy shares (26%, 37%, 53%, 61%, 67%, and 85%). In-cylinder pressure decreases with increasing CNG energy share (Figure 8.3a). The start of combustion was slowed with an increasing energy share due to a higher ignition delay because of burning of low reactivity fuel (CNG) (Figure 8.3b). Peak heat release rate is higher with CNG energy shares (26%, 37%, and 53%) than that of base diesel. However, it is lower with other higher energy shares. The in-cylinder temperature is almost the same with CNG energy shares (26% and 37%) compared to base diesel. However, it is lower with an 85% CNG energy share. Figure 8.4a, b, and c indicates variation of in-cylinder heat release rate and in-cylinder temperature with respect to crank angle in the engine in an RCCI mode (diesel-HCNG). The peak heat release rate is higher with HCNG than CNG energy share. The other results are the same for both CNG and HCNG fuel.

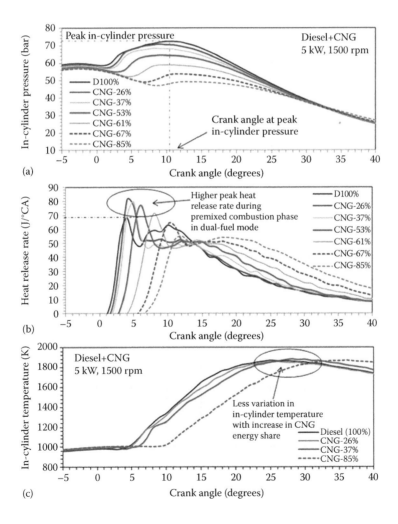

FIGURE 8.3
(a) In-cylinder pressure, (b) heat release rate, and (c) in-cylinder temperature of a diesel-CNG-fueled diesel engine at 5-kW power output. (From Singh, S. (2017), Utilization of methane in a compression-ignition engine under dual fuel mode, PhD thesis, IIT Delhi.)

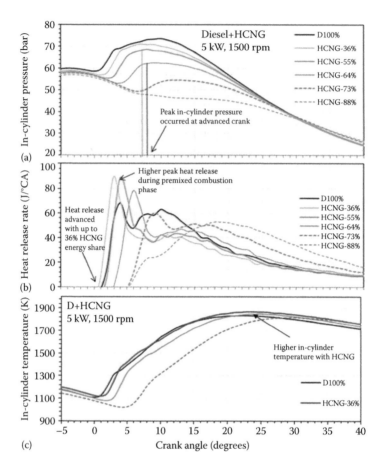

FIGURE 8.4

(a) In-cylinder pressure, (b) heat release rate, and (c) in-cylinder temperature of a diesel-HCNG-fueled diesel engine at 5-kW power output. (From Singh, S. (2017), Utilization of methane in a compression-ignition engine under dual fuel mode, PhD thesis, IIT Delhi.)

Figure 8.5 shows variation of HC, CO, NOx, and smoke emissions of a diesel engine (2 and 5 kW) in an RCCI mode (diesel-different gaseous fuels (CNG/HCNG)). NOx emission decreases with increasing energy shares (CNG/HCNG) compared to base diesel mode. The NOx emission at high power output (5 kW) is higher than low power output (2 kW). The main reasons for less NOx emission include these gaseous fuels having a higher specific heat of a charge, a reduction in oxygen concentration, and diluting the charge. If NOx emission with HCNG is compared with CNG, the emission is higher with HCNG than CNG due to the high flame velocity of hydrogen and hence a higher in-cylinder temperature. However, note that NOx emission with both CNG and HCNG energy shares is less than base diesel.

Smoke emission decreased drastically with HCNG and CNG energy shares compared to base diesel (Figure 8.5). This is primarily due to an increase in the premixed charge as the gaseous fuel is inducted along with air during a suction stroke, resulting in an increased premixed charge. The smoke emission at both power outputs (2 and 5 kW) is still less with HCNG than with CNG energy shares. However, hydrocarbon and carbon monoxide emissions are higher with both CNG/HCNG energy shares compared to base diesel. These emissions are higher at lower power output (2 kW) than higher power output (5 kW). The reasons

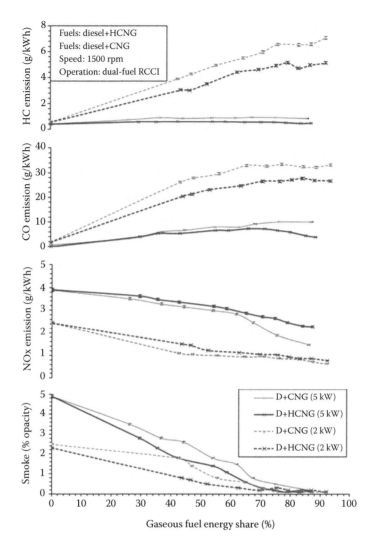

FIGURE 8.5
HC, CO, NOx and smoke emissions of diesel engine using different gaseous fuels. (From Singh, S. (2017), Utilization of methane in a compression-ignition engine under dual fuel mode, PhD thesis, IIT Delhi.)

for the increase of the emissions are due to the narrow flammability limit of the CNG/HCNG share (Liao et al., 2005) especially at low load, and trapped gaseous fuel in the crevice volume. Lower in-cylinder temperature at lower output is also the main reason for higher CO and HC emissions as these are unable to oxidize into CO_2 at low temperatures. These emissions can be reduced significantly with hydrogen-blended CNG but these emissions are still higher than base diesel. Overall, utilization of high-octane-number gaseous fuels (CNG and HCNG) in diesel engine in an RCCI mode could lead to high thermal efficiency of the engine along with lower NOx and smoke emissions. However, HC and CO emissions are higher at part load. These emissions could be reduced using exhaust gas recirculation and aftertreatment devices. HC and CO emissions can effectively be reduced using the available technology but reduction of NOx and PM emissions is difficult. In this context, NOx and PM emissions can be significantly reduced in compression-ignition engines using RCCI technology.

Summary

- Higher-octane-number fuels such as gasoline, methane, ethanaol, and methanol, which are called low reactivity fuels, can be used in compression igntion engines in an RCCI mode.
- Higher-cetane-number fuels, which are called high reactivity fuels, are used as ignition sources for combusting low reactivity fuels. The combustion is controlled by changing the reactivity of the fuels.
- NOx and PM emissions in CI engines in an RCCI mode decreased significantly.
- Thermal effiency could be increased in an RCCI mode.
- HC and CO emissions at high load incresed moderately but are higher at low or part loads.
- Emissions in an RCCI mode decreased with three orders of magnitude of NOx, six orders of magnitude of soot, and an increase of 16.4% indicated efficiency compared to diesel combusiton in conventional compression igntion engines (Kokjohn et al., 2011).
- Reduction of NOx and PM emission at the source level in compression-ignition engines is possible in an RCCI mode.

It can also be expected that the use of both high reactivity biofuels (biodiesel, F-T diesel, and DME) and low reactivity biofuels in compression-ignition engines under RCCI mode could yield beneficial results. In order to confirm the benefits, more studies focusing on RCCI with biofuels need to be conducted.

Solved Numerical Problem

1. A compression-ignition engine operates with methane and biodiesel in a dual-fuel mode. The mass flow rate of methane and biodiesel is 0.3 and 0.6 kg/s, respectively. The calorific value of methane and biodiesel is 48 and 38 MJ/kg, respectively. Thirty percent of biodiesel fuel is injected before the start of combustion and the rest is injected after the start of combustion. The global equivalence ratio of air-fuel (methane + biodiesel) is 0.8:1. Calculate the following parameters:

   a. Premixed charge ratio (mass and energy basis)

   b. Energy share of methane

   Input Data

   Input parameters:

   Mass flow rate of methane (CH_4) = $\dot{m}_{CH4}$ (kg/s) = 0.3 kg/s

   Mass flow rate of biodiesel ($C_{18}H_{22}O_3$) = $\dot{m}_{B100}$ (kg/s) = 0.6 kg/s

   Calorific value of CH_4 = CV_{CH4} (MJ/kg) = 48 MJ/kg

   Calorific value of biodiesel = CV_{B100} (MJ/kg) = 38 MJ/kg

   Biodiesel injected prior to start of combustion = $\dot{m}_{B30}$ (kg/s) = 30% total mass of B100

   $\dot{m}_{B30}$ (kg/s) = 0.3 × 0.6 = 0.18 kg/s

Solution

a. Premixed charge ratio on energy basis could be determined using.

$$PC_{energy\ ratio} = \left(\frac{\dot{m}_{CH_4} \times CV_{CH_4} + \dot{m}_{B30} \times CV_{B100}}{\dot{m}_{CH_4} \times CV_{CH_4} + \dot{m}_{B100} \times CV_{B100}} \right) 100(\%)$$ (8.1)

$$PC_{energy\ ratio} = \left(\frac{0.3 \times 48 + 0.18 \times 38}{0.3 \times 48 + 0.6 \times 38} \right) 100(\%)$$

$$PC_{energy\ ratio} = 57\%$$

Premixed charge ratio on mass basis could be determined using

$$PC_{mass\ ratio} = \left(\frac{\dot{m}_{CH4} + \dot{m}_{B30}}{\dot{m}_{CH_4G} \times CV_{CH_4} + \dot{m}_{B100} \times CV_{B100}} \right) 100(\%)$$ (8.2)

$$PC_{mass\ ratio} = \left(\frac{0.3 + 0.18}{0.3 + 0.6} \right) 100(\%)$$

$$PC_{mass\ ratio} = 53.3\%$$

b. Energy share of methane could be determined using

$$E_{CH_4} = \left(\frac{\dot{m}_{CH_4} \cdot CV_{CH_4}}{\dot{m}_{CH_4} \times CV_{CH_4} + \dot{m}_{B100} \times CV_{B100}} \right) 100(\%)$$ (8.3)

$$E_{CH_4} = \left(\frac{0.3 \times 48}{0.3 \times 48 + 0.6 \times 38} \right) 100(\%)$$

$$E_{CH_4} = 38.7\%$$

References

Aceves, S. M., Smith, J. R., Westbrook, C.K. and Pitz, W. J. (1999), Compression ratio effect on methane HCCI combustion, *Journal of Engineering for Gas Turbines and Power, Transactions of the ASME*, 121(3), 569–574.

Au, M. Y., Girard, J. W., Dibble, R., Flowers, D. et al. (1999), 1.9-liter four-cylinder HCCI engine operation with exhaust gas recirculation, SAE Technical Paper, 2001-01-1894 (1).

Banapurmath, N. R., Tewari, P. G. and Hosmath, R. S. (2008), Experimental investigations of a four-stroke single cylinder direct injection diesel engine operated on dual fuel mode with producer gas as inducted fuel and honge oil and its methyl ester (HOME) as injected fuels, *Renewable Energy*, 33(9), 2007–2018.

Banapurmath, N. R., Tewari, P. G., Yaliwal, V. S. et al. (2009), Combustion characteristics of a 4-stroke CI engine operated on honge oil, neem and rice bran oils when directly injected and dual fuelled with producer gas induction, *Renewable Energy*, 34(7), 1877–1884.

Bedoya, I. D., Saxena, S., Cadavid, F. J., Dibble, R. W. et al. (2012), Experimental study of biogas combustion in an HCCI engine for power generation with high indicated efficiency and ultra-low NOx emissions, *Energy Conversion and Management*, 53(1), 154–162.

Chen, Z., Wu, Z., Liu, J. and Lee, C. (2014), Combustion and emissions characteristics of high nbutanol/diesel ratio blend in a heavy-duty diesel engine and EGR impact, *Energy Conversion and Management*, 78, 787–795.

Chintala, V. and Subramanain, K. A. (2015), Experimental investigations on effect of different compression ratios on enhancement of maximum hydrogen energy share in a compression ignition engine under dual-fuel mode, *Energy*, 87, 448–462.

Dempsey, A. B. and Reitz, R. D. (2011), Computational optimization of a heavy-duty compression ignition engine fuelled with conventional gasoline, SAE Technical Paper 2011-01-0356.

Dempsey, A., Walker, N., Gingrich, E. and Reitz, R. D. (2014), Comparison of low temperature combustion strategies for advanced compression ignition engines with a focus on controllability, *Combustion Science and Technology*, 86(2), 210–241.

Geo, V. E., Nagarajan, G. and Nagalingam, B. (2008), Studies on dual fuel operation of rubber seed oil and its bio-diesel with hydrogen as the inducted fuel, *International Journal of Hydrogen Energy*, 33(21), 6357–6367.

Goran, H., Hyvonen, J., Tunestal, P. and Johansson, B. (2004), HCCI closed-loop combustion control using fast thermal management, SAE Technical Paper 2004-01-0943.

Hardy, W. L. and Reitz, R. D. (2006), A study of the effects of high EGR, high equivalence ratio, and mixing time on emissions levels in a heavy-duty diesel engine for PCCI combustion, SAE Technical Paper 2006-01-0026.

Helmantel, A. and Denbratt, I. (2004), HCCI operation of a passenger car common rail DI diesel engine with early injection of conventional diesel fuel, SAE International Technical Paper, 2004-01-0935.

Homan, H. S., Reynolds, R. K., De Boer, P. C. T. and McLean, W. J. (1979), Hydrogen-fueled diesel engine without timed ignition, *International Journal of Hydrogen Energy*, 4(4), 315–325.

Hunter, M. J., Flowers, D. L., Buchholz, B. A. and Dibble, R. W. (2005), Investigation of HCCI combustion of diethyl ether and ethanol mixtures using carbon 14 tracing and numerical simulations, *Proceedings of the Combustion Institute*, 30, 2693–2700.

Ibrahim, M. M. and Ramesh, A. (2013), Experimental investigations on a hydrogen diesel homogeneous charge compression ignition engine with exhaust gas recirculation, *International Journal of Hydrogen Energy*, 38(24), 10116–10125.

Inagaki, K., Fuyuto, T., Nishikawa, K., Nakakita, K. et al. (2006), Dual-fuel PCI combustion controlled by in-cylinder stratification of ignitability, SAE Technical Paper 2006-01-0028.

Kalsi, S. S. and Subramanian, K. A. (2016), Experimental investigations of effects of EGR on performance and emissions characteristics of CNG fuelled reactivity controlled compression ignition (RCCI) engine, *Energy Conversion and Management*, 130, 91–105.

Kokjohn, S., Hanson, R., Splitter, D. and Reitz, R. (2011), Fuel reactivity controlled compression ignition (RCCI): A pathway to controlled high-efficiency clean combustion, *International Journal of Engine Research*, 12(3), 209–226.

Korakianitis, T., Namasivayam, A. M. and Crookes, R. J (2010), Hydrogen dual-fuelling of compression ignition engines with emulsified biodiesel as pilot fuel, *International Journal of Hydrogen Energy*, 35(24), 13329–13344.

Li, Y., Jia, M., Chang, Y., Liu, Y. et al. (2014), Parametric study and optimization of a RCCI (reactivity controlled compression ignition) engine fueled with methanol and diesel, *Energy*, 65, 319–332.

Liao, S. Y., Cheng, Q., Jiang, D. M. et al. (2005), Experimental study of flammability limits of natural gas–air mixture, *Hazardous Materials*, 119, 81–84.

Noehre, C., Anderson, M., Johnson, B. and Hultqvist, A. (2006), Characterization of partially premixed combustion, SAE Technical Paper 2006-01-3412.

Opat, R., Ra, Y., Gonzalez, M. A., Krieger, R. et al. (2007), Investigation of mixing and temperature effects on HC/CO emissions for highly dilute low temperature combustion in a light duty diesel engine, SAE Technical Paper 2007-01-0193.

Reitz, R. D. and Duraisamy, G. (2015), Review of high efficiency and clean reactivity controlled compression ignition (RCCI) combustion in internal combustion engines, *Progress in Energy and Combustion Science*, 46, 12–71.

Ryu, K. (2013), Effects of pilot injection pressure on the combustion and emissions characteristics in a diesel engine using biodiesel–CNG dual fuel, *Energy Conversion and Management*, 76, 506–516.

Saxena, S. and Bedoya, I. D. (2013), Fundamental phenomena affecting low temperature combustion and HCCI engines, high load limits and strategies for extending these limits, *Progress in Energy Combustion Science*, 39, 457–488.

Sayin, C., Ilhan, M., Canakci, M. and Gumus, M. (2009), Effect of injection timing on the exhaust emissions of a diesel engine using diesel–methanol blends, *Renewable Energy*, 34(5), 1261–6129.

Singh, S. (2017), Utilization of methane in a compression ignition engine under dual fuel mode, PhD thesis, IIT Delhi.

Splitter, D., Hanson, R, Kokjohn, S. and Reitz, R. (2011), Reactivity controlled compression ignition (RCCI) heavy-duty engine operation at mid-and high-loads with conventional and alternative fuels, SAE Technical Paper 2011-01-0363.

Splitter, D., Kokjohn, S., Rein, K., Hanson, R. et al. (2010), An optical investigation of ignition processes in fuel reactivity controlled PCCI combustion, SAE Technical Paper 2010-01-0345.

Subramanian, K.A. and Chintala, V. (2013), Reduction of GHGs emissions in a biodiesel fuelled diesel engine using hydrogen in ASME 2013 Internal Combustion Engine Fall Technical Conference, Dearborn, Michigan, Technical Paper No. ICEF2013-19133, doi:10.1115/ICEF2013-19133.

Swami Nathan, S., Mallikarjuna, J. M. and Ramesh, A. (2010), An experimental study of the biogas–diesel HCCI mode of engine operation, *Energy Conversion and Management*, 51(7), 1347–1353.

Takahashi, S. (1982), An experiment on the ignition of hydrogen injected into a high temperature oxidizer, *International Journal of Hydrogen Energy*, 7(7), 589–596.

Tsolakis, A., Hernandez, J. J., Megaritis, A. and Crampton, M. (2005), Dual fuel diesel engine operation using H_2. Effect on particulate emissions, *Energy & Fuels*, 19(2), 418–425.

Turns, S. R. (1999), *An Introduction to Combustion: Concepts and Applications*, Third Edition, WCB/McGraw-Hill, New York.

Wu, H.-W. and Wu, Z.-Y. (2013), Using Taguchi method on combustion performance of a diesel engine with diesel/biodiesel blend and port-inducting H_2, *Applied Energy*, 104, 362–370.

Yao, M., Zheng, Z. and Liu, H. (2009), Progress and recent trends in homogeneous charge compression ignition (HCCI) engines, *Progress in Energy Combustion Science*, 35(5), 398–437.

9

Effect of Biofuels on GHGs

9.1 Introduction to GHGs

In 1827, Joseph Fourier, a French mathematician and physicist, investigated why Earth's average temperature is approximately 15°C (59°F), as his calculations indicated that the average temperature of Earth should actually be much colder—about –18°C or 0°F. He explained that there must be some type of balance between the incoming energy and the outgoing energy to maintain this constant temperature. The Intergovernmental Panel on Climate Change (IPCC) forecasts Earth's temperature, which will rise in a range from 2.5°F to 10°F in the next century due to an increase in concentrations of GHG emissions (NASA). In 1896, Svante Arrhenius published a paper in *Philosophical Magazine and Journal of Science* and reported how carbon dioxide contributes to the greenhouse effect. He speculated that carbon dioxide warms Earth by trapping heat near the surface. Today the world is experiencing first-hand the link between and the negative consequences of burning fossil fuels and global warming in the form of extreme climate changes, including glacier melting, ocean level rise, abnormal flood and drought, ocean acidification, sea level rise, hurricanes, tornadoes, loss of some habitats, extinction of some species/microorganisms, and detriment of agricultural productivity, all of which may likely result in severe damage to the ecosystem.

9.1.1 Leading GHGs

A gas that contributes to the greenhouse effect by absorbing infrared radiation is called a greenhouse gas (GHG). The IPCC lists the following gases that contribute to the greenhouse effect. These gases trap the heat from sunlight in the atmosphere and make the planet warmer. The primary sources of GHGs are electricity production, transportation, industry, commercial and residential buildings, agriculture, and land use and forestry.

- Carbon dioxide (CO_2)
- Methane (CH_4)
- Nitrous oxide (N_2O)
- Chlorofluorocarbons (CFC-11, CFC-12, CFC-113)
- Hydrofluorocarbons (HCFC-22, HCFC-141b, HCFC-142b)
- Methyl chloroform (CH_3CCl_3)
- Carbon tetrachloride (CCl_4)
- Pentafluoroethane (HFC-125)

- Tetraflouroethane (HFC-134a)
- Difluoroethane (HFC-152a)
- Fluoroform (HFC-223)
- Sulfur hexafluoride (SF-6)
- Tetrafluoramethane (CF_4 (PFC-14))
- Hexafluorethane (C_2F_6 (PFC-116))

9.2 Greenhouse Effect from GHGs in the Atmosphere

Earth receives energy that travels from the sun in a variety of wavelengths, including shorter-wavelength ultraviolet radiation and longer-wavelength infrared radiation. Some of the solar energy is absorbed by Earth itself through GHGs present in the atmosphere, leading to warming of the planet's surface. The wavelength of CO_2, CH_4, and N_2O match with certain wavelengths of the sun's radiation, and these gases trap the sun's energy, resulting in warming of Earth's atmosphere. The warming of Earth affects the lithosphere and hydrosphere.

Emissions such as CO_2, CH_4, and N_2O are considered GHGs because they trap sun heat and retain it, resulting in warming up of the atmosphere. Water vapor in the atmosphere is the most abundant heat-trapping gas but it is not considered a GHG because it has a short cycle (evaporation, lifetime, and condensation) in the atmosphere (a few days). The effect of water vapor on the greenhouse effect is much less compared to other GHGs such as CO_2.

The lifetimes of GHGs vary. For example, the lifetime of N_2O is higher than those of CH_4 and CO_2 emissions. The GWP is a relative measure of how much heat a GHG traps in the atmosphere that is based on its greenhouse effect for a specific time interval of 20 years, 100 years, and 500 years. The GWP for CO_2, CH, and N_2O for a 20-year horizon is 1, 72, and 289, respectively (IPCC, 2007). The GWP is calculated based on the ratio of the warming of the atmosphere caused by one substance (CH_4, N_2O, etc.) to that caused by a similar mass of carbon dioxide. The GWP of CO_2 emission is such a substance and is used as the reference for calculating the GWP of other GHG gases. For example, the GWP of N_2O is 289, which means the greenhouse effect of N_2O is 289 times higher than that of CO_2 emission if both species (CO_2 and N_2O) are compared at the same mass. It can be interpreted from the GWP that 1 g of N_2O emission in the atmosphere is almost equal to the effect of 289 g of CO_2. Similarly, methane gas also has a higher GWP than CO_2. The CO_2 emission from internal combustion engines is larger than that of CH_4 and N_2O emissions. CO_2 emission is described in terms of percentage of exhaust gas, whereas CH_4 is described in terms of parts per million (ppm), and N_2O is described in parts per billion (ppb). Even though CH_4 and N_2O are described in ppm/ppb levels, their greenhouse effect is higher and needs to be taken in account for calculating total CO_2-equivalent GHG emissions.

The Environmental Protection Agency (EPA) reported that GHG emissions from the worldwide transportation sector are about 13% (EPA), but in 2012 these emissions were still higher in the United States (28%), the European Union (23%), and the United Kingdom (20%) (Department of Energy & Climate Change, UK).

Global CO_2 emissions from fossil fuel combustion and some industrial processes in 2011 are shown in Figure 9.1 (EPA, 2014). Apportionment of the percentage of the CO_2 emission in China, the United States, and India were 28%, 16%, and 6%, respectively. These emissions can be linked with economic growth of the countries. CO_2 emission from the

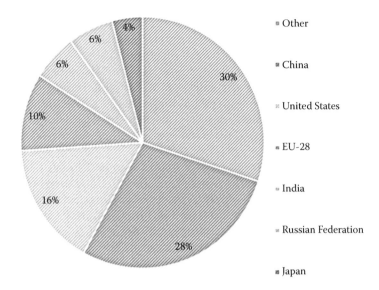

FIGURE 9.1
Global CO_2 emissions from fossil fuel combustion and some industrial processes in 2011. From United States Environmental Protection Agency, Global greenhouse gas emissions data. http://www.epa.gov/climatechange /ghgemissions/global.html, accessed April 6, 2014.)

transport sector for 2030 is projected to be 1278 million tonnes (China), 455 million tonnes (India), and 1840 million tonnes (United States) (IEA, 2008).

9.3 GHG Emissions from Internal Combustion Engines

Fuels for internal combustion engines are mostly hydrocarbons (diesel, gasoline, propane, butane, methane, biogas, producer gas, biodiesel, FF-T diesel, DME, etc.). The air (1 mole of oxygen and 3.76 moles of nitrogen in the atmosphere) is used as a working fluid and oxidizer in internal combustion engines. A chemical reaction occurs between hydrocarbon fuel and oxygen during combustion, through which carbon in the fuel converts into CO_2 through intermediate products such as CO, UHC, partially burned hydrocarbon, CH_4, soot, and PM, whereas hydrogen in fuel is converted into water vapor through intermediate products such as hydrogen and radicals. In this combustion process, nitrogen in the air reacting with oxygen at high temperature forms NO, NO_2, and N_2O through a complex chemical reaction. In the case of hydrogen, no carbon-based emissions (CO, HC, CO_2, soot, etc.) would come from the engines; however, NO, NO_2, and N_2O would form. Therefore, hydrogen-fueled engines cannot be described as having completely GHG-free emissions. CO_2, CH_4, and N_2O gases are considered GHGs because they trap the sun's radiation in their wavelengths, which fall within the infrared wavelength range of the sun's radiation. Therefore, the following three GHGs are explained in the following sections in detail:

- CO_2
- CH_4
- N_2O

CO_2 emission from the transportation sector can be reduced using different control strategies, including implementation of fuel economy standards and emission norms. For example, the European Union has decided to implement CO_2 emission norms (120 g/km) for its passenger vehicles. The EPA and the National Highway Traffic Safety Administration (NHTSA) developed the GHGs emission standards (CO_2: 600 g/bhp-hr, CH_4: 0.1 g/bhp-hr, and N_2O: 0.1 g/bhp-hr) for CI engines (Federal Register, 2011). Even though these regulations are mainly targeted for control of CO_2 emission, they do not address the control strategy for N_2O and CH_4 emissions from the transportation sectors as these emissions have higher GWP than CO_2.

Many organizations, including the National Action Plan on Climate Change (NAPCC), the EPA, and the NHTSA have initiated certain measures for mitigation of GHG emissions. The NAPCC suggested two major initiatives for emissions reduction: energy efficiency improvement and the utilization of alternative fuels in IC engines. Control strategies for GHG emission mitigation includes thermal efficiency improvement, use of alternative fuels, improvement in traffic management systems, and system weight reduction (Subramanian et al., 2008). The research findings of Subramanian and Chintala revealed the use of gaseous fuels (CNG, hydrogen) in a CI engine (7.4 kW rated power at 1500 rpm) in a dual-fuel mode could reduce engine-out GHG emissions and improve the energy efficiency of the engine significantly (Subramanian et al., 2013).

9.4 Are Biofuels Carbon-Neutral?

If the mass of carbon fixation by a bioplant is the same as released back into the atmosphere, net carbon is zero and it can be called a carbon-neutral fuel. This process can be described in many ways, as follows.

Carbon-neutral means that producing and burning a carbon-based fuel will not increase the carbon dioxide in the atmosphere. Carbon-neutral can be defined as when the amount of carbon dioxide released into the atmosphere is the same as the amount absorbed by the plants while growing. For example, consider a tree growing naturally in a forest and then burned completely due to a natural forest fire. Carbon from the atmosphere was converted by the tree through the photosynthesis process called carbon fixation and the carbon was released back into the atmosphere. Hence, there is no new addition of CO_2 in the atmosphere, which is called carbon-neutral. As another example, consider a tree growing in a garden with artificial nutrients (urea and other supplements). People were needed to do the watering. The nutrients that were produced through the fossil fuel- (natural gas) based fertilizer industry were used for better growth of the tree. The tree was then cut using an electric hacksaw powered by electricity generated by a fossil fuel (coal) power plant. Transportation of the tree was done by a fossil-fueled (diesel) truck. The tree was then accidently set on fire and released carbon back into the atmosphere. Could we say the tree is carbon-neutral? The answer is that based on the life cycle of the tree, it is not completely carbon-neutral because the embodied energy from a fossil fuel was used, which resulted in additional CO_2 emission released into the atmosphere. Carbon fixation by the tree is not equal to the release of it back into the atmosphere. In fact, the carbon released by the tree into the atmosphere is higher than the carbon fixation. Biofuels may produce less carbon dioxide, and therefore, they are not carbon-neutral because fossil fuels are used in their production, for example in making fertilizers for growing plants.

A system in a life cycle needs a great deal of energy for manufacturing of a product as well its operation. For example, the making of an automobile requires embodied energy for production and machining of components, assembly of the components, transporting of the vehicle to the retailer, and so forth. The vehicle itself needs fuel/energy to function. A system needs energy for it to be made, which is called embodied energy. Thus, for example, energy is consumed by a system to create a building, including extraction, processing and manufacture, and transportation.

Biomass needs embodied energy for its growth and energy for the production, storage, transportation, and dispensing of biofuels. The degree of being carbon-neutral may not be 100% for biofuels, and therefore, it needs to be studied in detail.

Well-to-wheel efficiency is used for GHG emission and energy analysis of the life cycle of fuels and vehicles. Well-to-tank efficiency indicates the energy input requirement needs for exploration of crude oil/biomass, production, storage, transportation, and dispensing. Tank-to-wheel efficiency indicates the efficiency of a vehicle on the road. For example, if a vehicle's tank-to-wheel efficiency is higher with hydrogen than with natural gas, the well-to-tank efficiency will be less with hydrogen than with natural gas since hydrogen needs more energy input for production and compressed or cryogenic storage than natural gas. The well-to-wheel efficiency of various vehicles fueled with different fuels are given in Table 9.1. Well-to-wheel analysis is associated with GHG emissions released during the system's life cycle. If a system needs more energy input with a high level of release of GHG emissions during its life cycle period, the system is viewed as less sustainable.

TABLE 9.1

Well-to-Tank, Tank-to-Wheel, and Well-to-Wheel Efficiency

Fuel	Type of Vehicle	Well-to-Tank Efficiency (%)	Tank-to-Wheel Efficiency (%)	Well-to-Wheel Efficiency (%)
Conventional gasoline	SI	89.6	12.1	10.6
Conventional diesel	CI	86	18.2	15.6
Natural gas	CNG (conventional)	69	26.1	18
Methanol	Fuel cell (fuel processor)	51	37.2	19
Liquid hydrogen	Fuel cell	57.7	22.5	13
Gasoline fuel cell hybrid vehicle	Fuel cell	88	16.2	14.2
Gasoline-fueled hybrid vehicle	Electric/battery	88	22.3	19.6
Diesel-fueled hybrid vehicle	Electric/battery	86	30.1	25.9

Source: Sheldon, A. E. and Williamson, S. (2005), Comparative assessment of hybrid electric and fuel cell vehicles based on comprehensive well-to-wheels efficiency analysis, *IEEE Transactions on Vehicular Technology*, 54(3), 856–862; Specht, B. A. (1998), Synthesis of methanol from biomass/CO_2 resources, 4th International Conference on Greenhouse Gas Control Technologies, Amsterdam; Troy, R. L. and Semelsberger, A. (2006), Dimethyl ether (DME) as an alternative fuel, *Journal of Power Sources*, 156(2), 497–511.

9.5 CO_2 Emission from Internal Combustion Engines

The stoichiometric chemical reaction for burning of hydrocarbon fuel is given in Equation 9.1. Hydrocarbon fuel burning with a mole of air would give b mole of carbon dioxide and c mole of water vapor, and $a3.76$ mole of nitrogen leaves intact. Carbon dioxide is the main product from combustion with hydrocarbon fuel.

$$C_xH_y + a(O_2 + 3.76N_2) \rightarrow bCO_2 + cH_2O + a3.76N_2 \tag{9.1}$$

The mass of carbon dioxide per kilogram of hydrocarbon is given in Table 9.2.

9.5.1 Carbon Balance Method for Determination of CO_2 Emission

Products from complete combustion in heat engines are carbon dioxide and water vapor. However, combustion in actual heat engines is incomplete and intermediate products are present in the exhaust gas, such as carbon monoxide, unburned hydrocarbon, methane, soot, and particulate matter. Therefore, the carbon in reactants (fuel) is equal to the carbon in products such as carbon monoxide, carbon dioxide, unburned hydrocarbon, soot, particulate matter. The carbon balance can be written as given in

$$C_xH_y + a(O_2 + 3.76N_2) \rightarrow bCO_2 + cH_2O + dCO + eC_xH_y + fCH_4$$
$$+ a3.76N_2 + gNO + hNO_2 + jN_2O \tag{9.2}$$

$$\sum_i^n C_{r,i} = \sum_i^n C_{p,i} \tag{9.3}$$

where
$C_{r,i}$ = Carbon in reactants from i.........n (C_xH_y, carbon content in other reactant, etc.)
$C_{p,i}$ = Carbon in products from i.........n (CO_2, CO, CH_4, PM, soot, etc.)

TABLE 9.2

Mass of Carbon Dioxide per Kilogram of Hydrocarbon

Hydrocarbons	Molecular Formula	Kilograms of CO_2/ Kilograms of Hydrocarbon Fuel
Methane	CH_4	2.75
Ethane	C_2H_6	2.93
Propane	C_3H_8	3
Butane	C_4H_{10}	3.03
Octane	C_8H_{18}	3.08
Decane	$C_{10}H_{22}$	3.09
Cetane	$C_{16}H_{34}$	3.11
Hydrogen	H_2	0

If combustion efficiency is 100%, the products are CO_2 and water vapor, and the carbon in the fuel is equal to the carbon in the CO_2. Equation 9.4 can be rewritten as given in Equation 9.5.

Carbon in Fuel = Carbon in CO_2

$$\frac{mw_c}{mw_f} \times \dot{m}_f = \frac{mw_c}{mw_{co_2}} \times \dot{m}_{co_2} \tag{9.4}$$

$$\dot{m}_{CO_2} = \frac{mw_{co_2}}{mw_f} \times \dot{m}_f \tag{9.5}$$

Equation 9.5 indicates that if the molecular weight of hydrocarbon decreases, then carbon dioxide emission increases but it is not correct. Because the energy content of fuel decreases with an increase in the molecular weight of hydrocarbon because the calorific value of pure substance of methane (CH_4) is less than propane, and similarly, butane is less than propane, and so forth. For example, if an engine is fueled with both methane and butane, the fuel flow rate of methane is not equal to that of butane as the calorific value of methane is higher than that of butane. The specific CO_2 emission per kilogram of hydrocarbon has already been discussed and shown in Table 9.2.

9.5.2 Relationship between CO_2 and Mass Flow Rate of Fuel

The mass of carbon dioxide emission from internal combustion engines is directly proportional to the mass flow rate of hydrocarbon fuel, as shown in Equation 9.6. The equation indicates that CO_2 emission increases with an increase in the mass flow rate of fuel.

$$\dot{m}_{CO_2} \, \alpha \, \dot{m}_f \tag{9.6}$$

9.5.3 Relationship between Mass Flow Rate of Fuel and Thermal Efficiency

If an internal combustion engine's brake power and a fuel's calorific value are kept constant, the mass of the fuel flow rate is indirectly proportional to the thermal efficiency of the engine, as shown in Equation 9.7. If the thermal efficiency of an engine increases, CO_2 emission decreases, and vice versa.

$$\dot{m}_f \, \alpha \, \frac{1}{\eta_t} \tag{9.7}$$

9.5.4 Relationship between CO_2 and Thermal Efficiency

The mass flow rate is the same in Equations 9.7 and 9.8; both equations can equate for showing the relationship between the mass of carbon dioxide and the thermal efficiency of internal combustion engines. Equation 9.8 indicates clearly that the mass flow rate of carbon dioxide is indirectly proportional to the thermal efficiency of engines. If the thermal efficiency of an engine increases, CO_2 emission decreases, and vice versa.

$$\dot{m}_{CO_2} \, \alpha \, \frac{1}{\eta_t} \tag{9.8}$$

It is clearly seen from the above discussion that the control measures of CO_2 emission in internal combustions engines are thermal efficiency improvement and use of fuel with less carbon content.

Electrical vehicles may emit zero emission but if the energy for battery charging comes from coal-based power plants, net CO_2 emission from electric vehicles will be higher than from conventional petroleum-fueled internal combustion vehicles. This occurs because coal-based power plants would have problems of losses through transmission and distribution but IC engines would have fewer losses.

9.5.5 Typical Example of Localized and Globalized Pollution

Localized and globalized pollution can be shown by comparing the overall efficiency of a battery-operated vehicle with a gasoline-fueled internal combustion vehicle.

Details of a battery operated vehicle:

Coal-based power plant efficiency = 35%

Transmission and distribution losses = 26%

Vehicle efficiency on the road = 75%

Battery charging and discharging efficiency = 80%

Overall efficiency of the complete system = $0.35 \times 0.74 \times 0.75 \times 0.80 = 15\%$

Diesel-fueled internal combustion efficiency:

Vehicle efficiency on road = 60% due to poor performance while idling and part load operation

Engine efficiency = 35%

Overall efficiency of the vehicle system = $0.6 \times 0.35 = 21\%$

The above example clearly shows that CO_2 emission from the overall electrical vehicle system is higher than from internal combustion vehicles. Even though localized pollution in an urban city will decrease, net emission will be higher with a battery-operated vehicle than with a conventional internal combustion engine. A battery-operated vehicle could provide better performance and emission benefits if the losses are addressed properly.

Similarly, it can be determined whether a natural gas fueled vehicle fleet is more advantageous compared with a hydrogen-fueled vehicle fleet.

For example, a gas refinery supplies the complete demand of natural gas to the transportation fleet in a city. As the air pollution level is high in the city, the administrator has taken a step to convert from a natural gas fleet to a hydrogen-fueled fleet. Subsequently, the refinery starts to produce hydrogen from natural gas using a steam reforming method.

The localized pollution is controlled in the city but net CO_2 emission is higher with hydrogen than with natural gas, because, the conversion efficiency of hydrogen to natural gas in the refinery is about 90%. One joule of natural gas energy is converted to 0.9 joule of hydrogen because conversion efficiency is less. Therefore, 1 joule of hydrogen energy needs 1.11 joules of natural gas energy input, and hence, the 11% additional increase in the quantity of natural gas has to be used for the production of hydrogen. Here, incremental costs for upgrading refineries, modification of infrastructure, and so forth, were not considered for this case and if these all were considered, a detailed analysis would required for addressing the air pollution issues in the city. If the hydrogen-fueled vehicle gives 11% higher efficiency, this exercise would yield beneficial results; otherwise, these types of mitigation measures may yield negative results and the base amount of air pollution in the city may be worsened further.

9.5.6 CO_2 Emission from Biofueled Engines

CO_2 emission from biofueled internal combustion engines is shown in Table 9.3. The change in CO_2 emission in an engine with base fuel to biofuel is in the range of 800 to 950 g/kWh. The data does not show whether the emission is higher or lower. Most of the CO_2 from biofueled engines could be recycled through a plant.

9.6 Effect of Gaseous Fuels (CNG/Hydrogen) on CO_2 Emission in a Dual-Fuel Diesel Engine (CNG-Diesel, Hydrogen-Diesel)

Some studies in the literature reported on the effect of gaseous fuels (CNG/hydrogen) on CO_2 emission in CI engines. For example, Korakianitis et al. reported decreasing CO_2 emission with a hydrogen-biodiesel-fueled engine as it decreased from 190 g/kWh in a base biodiesel mode to 104 g/kWh in a hydrogen-biodiesel dual-fuel mode (hydrogen consumption: 0.0005 kg/s) (Korakianitis et al., 2010). Similarly, Miyamoto et al. reported that CO_2 emission decreased from 688 g/kWh in a base diesel mode to 425 g/kWh in a hydrogen dual-fuel mode (10% hydrogen energy share) (Miyamoto et al., 2011). CO_2 emission decreased from 7.1% volume with base diesel to 4.3% volume with 65% CNG energy share in a CI engine (36 kW rated power at 3500 rpm) (Yusaf et al., 2010). Even though general information on CO_2 emission is available in the literature, specific information on the effects of energy efficiency and a fuel's carbon content on CO_2 emission is scant.

9.7 Effect of Carbon Content in Fuel on CO_2 Emission

Carbon content in fuel (stoichiometric/lean mixture) is generally oxidized during combustion and converted to CO and then to CO_2. The CO_2 emission decreases with a decrease in the carbon content of the fuel, which can be determined using Equation 9.9. The carbon content in the air-fuel charge was reduced by the addition of hydrogen/CNG to diesel,

TABLE 9.3

CO_2 Emission from Biofueled Internal Combustion Engines/Vehicles Compared to Base Fuel

Serial number	Engine Specifications	Fuels Used	CO_2 Emission	Base Fuel	CO_2 Emission	% Change	Reference
1	Lombardini, 3 LD 450, four-stroke, air-cooled diesel, single-cylinder, 8.1 kW at 3000 rpm, maximum torque (Nm) 28 Nm at 1800–2000 rpm	Soybean biodiesel	11.8%	Diesel	11.3%	−4.4	Can (2014)
2	Lombardini 6 LD 400, four-stroke, air-cooled diesel, single-cylinder, maximum torque 21 Nm at 2200 rpm	Apricot seed kernel oil, methyl ester	0.056 Kg/MJ	Diesel	0.06 Kg/MJ	6.7	Gumus et al. (2010)
3	Single-cylinder, water-cooled, direct injection	Polanga seed oil	0.6%	Diesel	0.6%	0.0	Sahoo et al. (2007)
4	Four-cylinder, water-cooled, swirl chamber, diesel engine	Waste-oil	3%	Diesel	2.6%	−15.4	Lin et al. (2014)
5	Kirloskar, single-cylinder, four-stroke, constant speed, diesel engine	Fish oil	0.7 g/kW-hr	Diesel	0.68 g/kW-hr	−2.9	Sakthivel et al. (2014)
6	6.0 L Ford Cargo, water-cooled, direct injection, naturally aspirated and four-stroke, in-line six-cylinder, 81 kW at 2600 rpm, 335 Nm at 1500 rpm	Canola oil methyl esters	10.5 g/kW-hr	Diesel	11.2 g/kW-hr	6.2	Ozsezen et al. (2011)
7	Six-cylinder, four-stroke, naturally aspirated, water-cooled diesel engine, rated power 205 kW, maximum torque 1100 Nm at 1300 rpm	Gas to liquid	0.8 kg/kW-hr	Diesel	0.78 kg/kW-hr	−2.6	Hassaneen et al. (2012)
8	Opel Rekord L, 4-cylinder, water-cooled, spark-ignition engine, maximum power 43 kW at 4300 rpm, maximum torque 125.6 Nm at 1800–2400 rpm	Ethanol + gasoline	12%	Gasoline	10.6%	−13.2	Yüksel et al. (2004)
9	Four-cylinder direct injection diesel engine, maximum torque 245 Nm at 1700 rpm, rated power 59 kW at 2800 rpm,			Ethanol	10.3%		He et al. (2003)

(Continued)

TABLE 9.3 (CONTINUED)

CO_2 Emission from Biofueled Internal Combustion Engines/Vehicles Compared to Base Fuel

Serial Number	Engine Specifications	Fuels Used	CO_2 Emission	Base Fuel	CO_2 Emission	% Change	Reference
10	6.0 L Ford Cargo, water-cooled, direct injection, naturally aspirated, four-stroke, in-line six-cylinder, 81 kW at 2600 rpm, 335 Nm at 1500 rpm	Palm oil methyl esters	11 g/kW-hr	Diesel	11.3 g/kW-hr	2.7	Ozsezen et al. (2009)
11	DM14, single-cylinder, naturally aspirated, power (HP) maximum 100, maximum speed range (rpm) 5650–8000	Jatropha	625 g/kW-hr	Diesel	589 g/kW-hr	−6.1	Raheman et al. (2014)
12	Rainbow-186 Diesel, single-cylinder, direct injection, maximum power 10 HP, maximum engine speed 3600 rpm	Waste anchovy fish biodiesel	2.55%	Diesel	3.5%	27.1	Behçet et al. (2011)
13	Six-cylinder, four-stroke, naturally aspirated, water-cooled diesel engine, rated power 205 kW, maximum torque 1100 Nm at 1300 rpm	Rapeseed methyl ester	0.77 kg/kW-hr	Diesel	0.8 kg/kW-hr	3.8	Ruschel et al. (2012)
14	Lombardini LM 250, four-stroke, single-cylinder, maximum speed (rpm) 3600, air- and water-cooled	Methanol	9.3%	Gasoline	13.7%	32.1	Celik et al. (2011)
15	Fiat-licensed Tofas 124, 4-cylinder, power 60 HP at 5600 rpm, torque 89 Nm at 3400 rpm	Hydrogen	0.001%	Gasoline	13%	100.0	Kahraman et al. (2007)
16	Proton Magma 12-valve, multicylinder, maximum output (DIN) PS/rpm net (kW/rpm) 87/6000 (64/6000), maximum torque (DIN) kg-m/rpm net (Nm/rpm) 12.5/3500 (122/3500)	CNG	8%	Gasoline	12%	33.3	Aslam et al. (2006)
17	Four-stroke, single-cylinder, spark ignition air-cooled engine	LPG	14%	Gasoline	14.8%	5.4	Çinar et al. (2016)
18	AKSA, A4CRX46TI, four-cylinder, constant speed engine	Plastic pyrolysis oil	11.5%	Diesel	10.4%	−10.6	Kalargaris et al. (2017)
19	SV1, Kirloskar oil engine limited, vertical, four-stroke, single-cylinder, rated output as per IS: 11170 8 HP (5.9 kW)	Annona biodiesel	6.8%	Diesel	7%	2.9	Ramalingam et al. (2016)

leading to a reduction in CO_2 emission of the engine. The CO_2 emission in a CI engine with stoichiometric air-fuel mixture is 3115 g CO_2/kg of diesel fuel, 3107 g CO_2/kg of diesel-CNG fuel, and 3103 g CO_2/kg of diesel-hydrogen fuel (Subramanian et al., 2013).

$$\text{Carbon content (\% by weight)} = \frac{\sum (n_{c,fi} \times MW_{c,fi})}{\sum (n_{c,fi} \times MW_{c,fi} + n_{H,fi} \times MW_{H,fi} + n_{O,fi} \times MW_{O,fi})} \times 100$$

$$(9.9)$$

9.8 Effect of Energy Efficiency on CO_2 Emission

Experimental work was conducted on a diesel engine in a dual-fuel mode (diesel-CNG and diesel-hydrogen) at 50% load and 100% load. Figure 9.2 shows the variation of CO_2 emission with respect to an increase in the gaseous fuel energy share at 50% and 100% engine loads. Energy efficiency is defined as the ratio of power output and heat energy input. For the same power output, the mass flow rate of fuel decreases with an increase in energy efficiency, resulting in CO_2 emission reduction. At a 19% gaseous fuel (CNG/hydrogen) energy share, CO_2 emission decreased about 40% and 28% in diesel-hydrogen and diesel-CNG dual-fuel modes, respectively. The CO_2 emission decreased from 948 g/kWh with base diesel to 446 g/kWh with a 47% CNG energy share. These results are in line with those reported by Lounici et al. that the CO_2 emission decreased from 8.9% in a base diesel mode to 6.9% in a CNG dual-fuel mode at 80% load and 1500 rpm (Lounici et al., 2014).

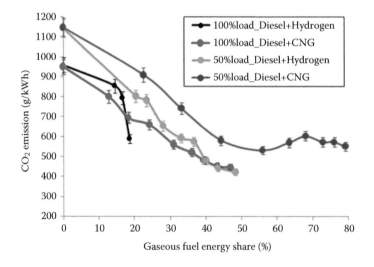

FIGURE 9.2
CO_2 emission variation with respect to gaseous fuel energy share. (From Chintala, V. (2016), Experimental investigation on utilization of hydrogen in a compression-ignition engine under dual fuel mode, PhD thesis, IIT Delhi.)

9.9 Methane Emission from Internal Combustion Engines

9.9.1 Methane Concentration in the Troposphere and Stratosphere

Methane (CH_4) emission is one of the GHG emissions that has a higher GWP of 25 (for a 100-year time horizon) than CO_2 emission. Methane is formed in the troposphere over the Earth's surface and it subsequently rises to the stratosphere, where it reacts with the hydroxyl radical (OH), as given in Equation 9.10, and CH_3 and water are products from this reaction. The oxidation of methane is the main source of water vapor in the upper stratosphere.

$$CH_4 + OH \rightarrow CH_3 + H_2O \qquad (9.10)$$

9.9.2 Methane, Nonmethane Hydrocarbon, and Total Hydrocarbon

Methane as an emission comes out from hydrocarbon-fueled internal combustion engines. Methane is a species in THC emission, which includes unburned hydrocarbon and partially oxidized hydrocarbon with different hydrocarbon chains (C_1 to C_n). THC consists of NMHC and methane, as shown in Equation 9.11. NMHC emission is currently used for the certification of vehicles, and based on measured emissions of THC and CH_4, NMHC can be calculated using Equation 9.12. Methane emission can be measured using a nondispersive infrared analyzer (NDIR), whereas THC is measured using the flame ionization detection (FID).

$$THC = NMHC + CH_4 \qquad (9.11)$$

$$NMHC = THC - CH_4 \qquad (9.12)$$

9.9.3 Formation Mechanism of Methane in Internal Combustion Engines

In methane-fueled internal combustion engines, UHC as a methane emission is higher during starting of the engine to until its warm-up period. In addition, combustion with a heterogeneous mixture (e.g., compression-ignition engine) wherever an ultralean air-fuel mixture (beyond the flammability limit) presents in a combustion chamber may not be able to oxidize methane into CO and then CO_2. In the case of longer-chain hydrocarbon as a fuel, methane emission is less or negligible.

Methane forms in all hydrocarbon-fueled internal combustion engines. The hydrocarbon undergoes thermal pyrolysis during combustion and a longer-hydrocarbon chain is broken into relatively smaller molecules (e.g., C_8 to C_1 for octane fuel (C_8H_{18})), including the lowest carbon numbered hydrocarbon (CH_4). The fragmented different hydrocarbon molecule needs different activation energy for oxidizing the hydrocarbon into CO and CO_2. The in-cylinder temperature distribution during combustion in the combustion chamber is different from the cylinder axis to the cylinder wall. The oxidation process of the fragmented hydrocarbons would depend on the in-cylinder temperature, which is relatively higher in the combustion zone and the center of the vertical cylinder axis, whereas it is lower near the cylinder wall due to more heat transfer loss from the cylinder wall to the cooling fins or cooling fluid. If the fragmented hydrocarbon presents in a low in-cylinder temperature environment, it (CH_4) cannot be oxidized to CO and CO_2 and the activation energy requirement for methane is also relatively higher, resulting in the presence of methane emission.

Few studies are available on the formation of CH_4 emission in CI engines in a hydrogen/CNG-based dual-fuel mode. The research findings of Subramanian and Chintala reported that methane emission could be reduced about 22% with the use of hydrogen (20% energy share) in a biodiesel-fueled CI engine (Subramanian et al., 2013). Conversely, the opposite is true for CNG fuel, as Timothy et al. observed that CH_4 emission increased from 2 to 14 g/bhp-hr with the use of CNG in a multicylinder CI engine (Gatts et al., 2012). CH_4 emission formation in CI engines is mainly dependent on the in-cylinder temperature of the engine (Subramanian et al., 2013). CH_4 emission reduces at high temperatures as its reaction rate is an exponential function of the engine's in-cylinder temperature (Equation 9.13) (Heywood et al., 1988).

$$d\,[CH_4]/dt = 1.3 \times 10^8 \exp(-24358/T)\,[CH_4]^{-0.3}\,[O_2]^{1.3} \qquad (9.13)$$

9.9.4 Effect of Gaseous Fuels (CNG/Hydrogen) on CH_4 Emission in a Dual-Fuel Diesel Engine (CNG-Diesel, Hydrogen-Diesel)

Injected diesel fuel ($C_{16}H_{34}$) is generally broken into smaller molecules due to thermal pyrolysis during the combustion process. The CH_4 molecule, which is a type of smaller molecule, forms through the thermal pyrolysis process. Even though some CH_4 molecules oxidize into CO_2, some molecules are unable to oxidize due to a variety of reasons, including low temperature prevailing in some zones (nearby wall, piston) in the combustion chamber, the presence of an overlean and rich mixture in a few zones (beyond the flammability limit), and complex chemical kinetic reaction. The activation energy requirement for the CH_4 molecule is higher than for diesel fuel. The ignition energy requirement for hydrogen combustion is less, which would lead to an enhanced species oxidation rate resulting in better conversion of methane into CO_2. However, the formation mechanism is different for a CNG-fueled dual-fuel engine. CNG fuel is generally mixed along with inducted intake air and then enters into the engine cylinder during a suction stroke. The established fact is that a CI engine operates with a heterogeneous mixture and the localized distribution of air and fuel varies from spray zone to other zones near the cylinder wall and piston. The oxidation rate of intermediate species is mainly dependent on the zone's localized air-fuel distribution and in-cylinder temperature in the combustion chamber of the engine. This study establishes that the formation of CH_4 and then its oxidation into CO and CO_2 are mainly temperature-dependent. However, in the case of a CNG-fueled CI engine, the formation mechanism is different as methane in the charge (air-CNG-diesel) is more dominant than the in-cylinder temperature effect.

The methane (CH_4) emission increased in a diesel-CNG dual-fuel mode, whereas it decreased in a hydrogen dual-fuel mode, as shown in Figure 9.3. Methane emission formation is a function of the methane content in the total energy share and in-cylinder temperature. The in-cylinder temperature is lower with a higher CNG energy share due to delaying the start of combustion and a too-lean mixture (beyond the flammability limit) existing in a nearby wall, resulting in higher methane emission. In addition, the unburned fuel could also enhance the emission formation due to flame quenching (flame interaction with relatively less gas temperature and a solid wall) and unburned fuel trapped in a crevice volume. Methane emission increased from 1.81g/kWh with base diesel to 7.71 g/kWh with 47% CNG energy share at the rated load. However, it increased at 50% load

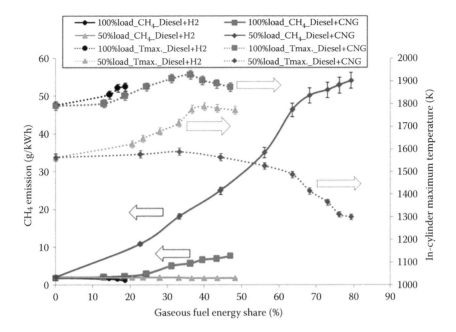

FIGURE 9.3
Effect of gaseous fuel addition on in-cylinder peak temperature and CH_4 emission.

(from 2.1 to 54.2 g/kWh) and the reason may be due to poor combustion with less quantity of pilot liquid fuel. The in-cylinder temperature decreased significantly with an increasing CNG share at 50% load. Lounici et al. reported similar results of a decrease in in-cylinder temperature at a lower load in a CI engine in a CNG-based dual-fuel mode due to charge dilution (Lounici et al., 2014). At lower load, Uma et al. also reported a similar trend of CH_4 emission that increased from 12.1 ppm in a base diesel mode to 32 ppm in a 40-kW CI engine (1500 rpm) in a dual-fuel mode (diesel-producer gas) (Uma et al., 2004).

Methane emission formation is lower in a hydrogen dual-fuel mode than in a CNG dual-fuel mode. At a 19% hydrogen energy share, the emission decreased about 40% in a hydrogen dual-fuel mode. This is mainly due to a significant increase in the in-cylinder temperature of the engine with the addition of hydrogen. A similar trend of increase in the in-cylinder temperature with a hydrogen addition of 7% volume in a CI engine was reported by Maghbouli et al. (2014). Bose et al. found a 3.2% increase in the in-cylinder temperature with the addition of hydrogen (0.15 kg/h) in a CI engine (5.2 kW rated power at 1500 rpm) in a dual-fuel mode (Bose et al., 2013). Hydrogen fuel has the desirable property of high flame speed, which leads to enhancement of the in-cylinder temperature and combustion rate, resulting in a substantial reduction of the CH_4 emission.

9.10 N_2O Emission from Internal Combustion Engines

Nitrous oxide (N_2O) emission is an important pollutant from the GHG family. Its lifetime is about 120 years, which means that its radioactive effects are continuous. N_2O emission generally forms by reaction with a fuel containing the nitrogen (N_2) compound and free radicals (O, H, and OH) under high temperature. Under high temperature, three body reactions can produce N_2O emission, as shown in

$$N_2 + O + M \rightarrow N_2O + M \tag{9.14}$$

Liu et al. reported various mechanisms for N_2O formation (NH_3 route and HCN route) in combustion of fossil fuel in a fluidized bed reactor and burner (Liu et al., 2002). In any combustion process, the N_2O emission forms at a temperature below 900°C (Federal Register, 2011). Most of the technologies, including EGR and aftertreatment devices (catalysts), focus on reducing NOx (NO + NO_2) emission in CI engines. However, these technologies may lead to increased N_2O emission. In a urea-based SCR system, 10% of NOx emission is converted to nitrous oxide emission due to a reduction in the catalyst temperature (Radojevic et al., 1998). Lu and Lu studied the influence of temperature (1073 K–1473 K) and O_2 concentration (0%–9.3%) on N_2O formation in a reactor (Lu et al., 2009). They concluded that N_2O emission decreases with a decrease in O_2 concentration. Caton studied the effect of oxygen concentration and temperature on N_2O emission from ammonia-based selective noncatalytic reactor (Caton et al., 1995). He reported that N_2O concentration decreased from 21 to 4 ppm with less O_2 concentration, but increased when the temperature increased from 1075 K to 1175 K (Caton et al., 1995).

In a CI engine, N_2O emission forms during combustion at a relatively lower in-cylinder temperature compared to the formation of NO and NO_2 emissions. However, it could also form at a high in-cylinder temperature and then it converts into NO/NO_2 emission. If the zone temperature is low, N_2O remains because it is unable to convert into NO. Therefore, N_2O emission formation is a temperature-dependent phenomenon. Figure 9.5 clearly indicates that N_2O emission is inversely proportional to the in-cylinder temperature. N_2O emission decreases drastically with an increase in in-cylinder temperature and vice versa. Peak in-cylinder temperature of the engine in a dual-fuel mode (diesel-hydrogen) increased steeply from 1793 K with base diesel to 1876 K with a 19% hydrogen energy share, and correspondingly, the N_2O emission decreased from 0.41 g/kWh in the base diesel mode to 0.32 g/kWh in the dual-fuel mode. N_2O emission is higher with 50% load (0.59 g/kWh) than 100% load (0.41 g/kWh) because peak in-cylinder temperature of the engine is lower with 50% load (1564 K) than 100% load (1793 K). In the case of a dual-fuel mode, at 50% load, N_2O emission is higher with an 80% CNG energy share (1.21 g/kWh) compared to a 30% CNG energy share (0.53 g/kWh) and base diesel (0.59 g/kWh) with the corresponding in-cylinder peak temperatures of 1301 K and 1589 K, respectively (Figures 9.4 and 9.5). A notable conclusion emerging from this study is that the N_2O emission is strongly dependent on in-cylinder temperature.

The above study results are in good agreement with the results of similar experimental works reported in the literature. For example, Feng et al. investigated the effect of temperature on reaction rate of N_2O emission in a fixed-bed reactor and reported that its reaction rate decreased from 17% to 8% when combustion temperature increased from 1023 K to 1123 K (Feng et al., 1996). Hayhurst and Lawrence also confirmed the same trend in coal combustion that N_2O emission decreased from 30 mgN_2O/MJ at 800°C to 7 mgN_2O/MJ at 900°C due to thermal decomposition of N_2O and its reaction with free radicals (O, OH, and H) (Hayhurst et al., 1992). It is clearly seen from Figure 9.10 that CI engine operation with low temperature leads to a high level of N_2O emission. At a 19% hydrogen energy share, N_2O emission decreased from 0.39 g/kWh in a base diesel mode to 0.29 g/kWh in a diesel-hydrogen dual-fuel mode. In a diesel-CNG dual-fuel mode, the emission decreased initially due to increase in combustion temperature but increased at a later stage due to decrease in the temperature as shown in Figure 9.10. A notable conclusion emerged from the study that

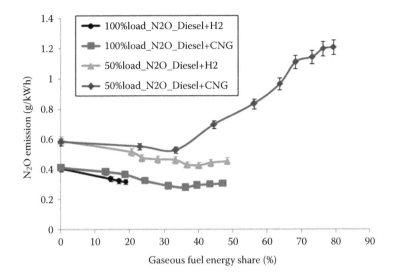

FIGURE 9.4
N$_2$O emission variation with respect to gaseous fuel energy share.

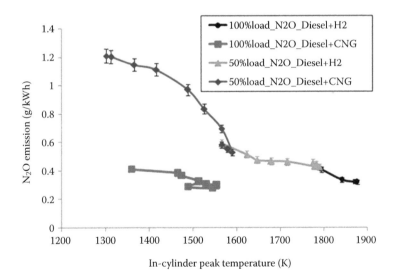

FIGURE 9.5
N$_2$O emission variation with respect to in-cylinder temperature.

N$_2$O emission is a strong function of in-cylinder temperature. The N$_2$O emission initially decreased from 0.39 g/kWh with base diesel fuel to 0.25 g/kWh with a 32% CNG energy share; however, it increased to 0.31 g/kWh with a 47% CNG energy share.

Technologies, including EGR and HCCI, which would bring down in-cylinder temperature (known as a low-temperature process), are solutions to NOx emission reduction in CI engines. However, these technologies may have a negative effect in terms of increasing N$_2$O emission.

The main GHGs, such as CO$_2$, N$_2$O, and CH$_4$ from base fuel (diesel/gasoline) and biofueled internal combustion engines/vehicles, are shown in Table 9.4. CO$_2$ emission is the

TABLE 9.4

GHG Emissions from Internal Combustion Engines/Vehicles

Serial Number	Engine Specifications	Fuels Used	CO$_2$	N$_2$O	CH$_4$	Reference
1	Euro 3 diesel Ford Connect TDCi with a manual 1.8 liter in-line four-cylinder, turbocharged, intercooled, eight-valve engine	Diesel	173.6 g/km	0.222 g/km	0.017 g/km	Przybyla et al. (2013)
2	Euro 3 diesel Ford Connect TDCi with a manual 1.8 liter in-line four-cylinder, turbocharged, intercooled, eight-valve engine	B100	172.9 g/km	0.023 g/km	0.012 g/km	Przybyla et al. (2013)
3	Kirloskar, EA10, single-cylinder, constant speed, rated output (kW) 7.4	B20 + H2 (20% energy share)	557.7 g/kW-hr	0.2 g/kW-hr	0.4 g/kW-hr	Subramanian et al. (2013)
4	Kirloskar, EA10, single-cylinder, constant speed, rated output (kW) 7.4	B20	1042 g/kW-hr	0.27 g/kW-hr	0.51 g/kW-hr	Subramanian et al. (2013)
5	2000 C15 Caterpillar engine, chassis dynamometer (Schenck-Pegasus unit) driven by a direct current (DC) 447 kW (600 HP) motor that can absorb up to 492 kW (660 HP).	Ultralow sulfur diesel	960 g/km	0.09 g/km	N/A	Na et al. (2015)
6	2000 C15 Caterpillar engine, chassis dynamometer (Schenck-Pegasus unit) driven by a direct current (DC) 447 kW (600 HP) motor that can absorb up to 492 kW(660 HP).	Soybean biodiesel (B100)	905 g/km	0.05 g/km	N/A	Na et al. (2015)
7	Ford Mondeo, multicylinder, 146 at 4500	Gasoline	260 g/km	0.001 g/km	0.02 g/km	Li et al. (2013)
8	2004 Cummins ISX450 engines, six-cylinder, 385–600 Hp, 287–447 kW, 2000–2100 rpm, 1966–2779 Nm	Diesel	1411 g/km	0.0966 g/km	0.076 g/km	Graham et al. (2008)
9	2004 Cummins ISX450 engines, six-cylinder, 385–600 Hp, 287–447 kW, 2000–2100 rpm, 1966–2779 Nm	CNG	1170 g/km	0.0580 g/km	6.26 g/km	Graham et al. (2008)
10	2004 Cummins ISX450 engines, 6-cylinder, 385–600 Hp, 287–447 kw, 2000–2100 rpm, 1966–2779 Nm	Hythane	1035 g/km	0.0348 g/km	5.03 g/km	Graham et.al (2008)
11	2004 Cummins ISX450 engines, six-cylinder, 385–600 Hp, 287–447 kW, 2000–2100 rpm, 1966–2779 Nm	Diesel	1631 g/km	0.0144 g/km	N/A	Graham et al. (2008)
12	2004 Cummins ISX450 engines, six-cylinder, 385–600 Hp, 287–447 kW, 2000–2100 rpm, 1966–2779 Nm	LNG + Diesel	1355 g/km	0.0204 g/km	2.62 g/km	Graham et al. (2008)

highest in terms of mass, whereas N_2O and CH_4 are the lowest. The CO_2 equivalent emission can be calculated using GWP factors.

Summary

- Carbon dioxide (CO_2) emission can be reduced in internal combustion engines by energy efficiency improvement and use of less or no carbon-content fuel. CO_2 emission from the biofueled engines could be recycled by the plants, resulting in no new addition of this emission into the atmosphere. Embodied energy and emissions would influence the degree of carbon-neutral fuel.

- Methane (CH_4) emission formation in internal combustion engines is mainly due to incomplete combustion. This emission is mainly a function of in-cylinder temperature. Biofuels used in internal combustion engines could reduce CH_4 along with CO, THC, and PM due to better combustion with these fuels, which have embedded oxygen.

- Nitrous oxide (N_2O) emission is an inverse function of the in-cylinder temperature of the engine. This emission is less dependent on the type of fuel (whether biofuel or fossil fuel) used in internal combustion engines.

Solved Numerical Problems

1. The mass flow rate of ethanol fuel in a spark-ignition engine is 12 kg/s. Let the engine operate with 100% combustion efficiency. Calculate the CO_2 emission rate of the engine.

   Input Data

   Mass flow rate of ethanol fuel (C_2H_5OH) = 2 kg/s

   Combustion efficiency = 100% (it means the product from the engine is only water vapor and CO_2)

   CO_2 emission rate (kg/s) = ?

   Solution

   $$C_2H_5OH + 3(O_2 + 3.76N_2) \rightarrow 2CO_2 + 3H_2O + 3.76N_2 \qquad (9.15)$$

   From Equation 9.15:

   $$CO_2 \text{ emission factor} = 2 \times \text{molecular weight of } CO_2$$
   $$/\text{molecular weight of } C_2H_5OH$$
   $$= 2 \times 44/46 = 1.91 \text{ kg of } CO_2/\text{kg of } C_2H_5OH$$

$$CO_2 \text{ emission rate} = CO_2 \text{ emission factor} \times \text{mass flow rate of } C_2H_5OH$$
$$\text{(for combustion efficiency: 100\%)}$$
$$= 1.91 \times 2 = 3.82 \text{ kg/s}$$

2. A butanol-fueled spark-ignition engine consumes fuel at rate of 5 kg/s. CO and HC emissions from the engine are 0.2 and 0.01 g/s, respectively. Calculate the CO_2 emission rate from the engine.

Input Data

Mass flow rate of ethanol fuel (C_2H_5OH) = 5 kg/s

Combustion efficiency is less than 100% as it means the product from the engine is not only water vapor and CO_2, but also CO and UHC (butanol)

CO_2 emission rate (kg/s) = ?

Solution

$$C_4H_9OH + a(O_2 + 3.76N_2) \rightarrow bCO_2 + cCO + dC_4H_9OH + c3H_2O + 3.76N_2 \quad (9.16)$$

From Equation 9.16:

The CO_2 emission factor can be calculated using the carbon balance method, as given below:

$$\text{Carbon in the fuel} = \text{Carbon in } CO_2 + \text{Carbon in } CO + \text{Carbon in UHC}$$

$$M_{C_4H_9OH} \times MW_C/MW_{C_4H_9OH} = M_{CO_2} \times MW_c/MW_{CO_2} +$$
$$M_{CO} \times MW_c/MW_{CO} + M_{C_4H_9OH} + MW_c/MW_{UHC}$$

$$M_{CO_2} = MW_{CO_2} \times (M_{C_4H_9OH}/MW_{C_4H_9OH} - M_{CO}/MW_{CO} - M_{C_4H_9OH}/MW_{UHC})$$

$$CO_2 \text{ emission rate} = 44 \times (5/74 - 0.2/28 - 0.01/74) = 2.65 \text{ kg of } CO_2$$

3. A diesel- ($C_{16}H_{34}$) fueled compression-ignition engine is to be converted to operate with DME (C_2H_5OH). The combustion efficiency, thermal efficiency, and power output of the engine with both fuels are the same. The lower calorific value of diesel and DME is 43 and 28 MJ/kg, respectively. Calculate the following parameters:

a. Specific molar concentration and mass of CO_2
b. Percent change in CO_2 emission
c. Specific carbon intensity
d. Specific CO_2 intensity

Input Data

Lower calorific value of diesel fuel = 43 MJ/kg

Lower calorific value of DME = 28 MJ/kg

Specific molar concentration and mass of CO_2 = ?

Specific carbon (kg/MJ) and CO_2 (kg/MJ) intensity = ?

$$C_{16}H_{34} + 24.5(O_2 + 3.76N_2) \rightarrow 16CO_2 + 17H_2O + 92.12N_2 \tag{9.17}$$

Specific CO_2 molar concentration = Number of CO_2 moles/Number of $C_{16}H_{34}$
= 16/1 = 16 moles of CO_2/1 mole of $C_{16}H_{34}$

Specific CO_2 mass emission = Mass of CO_2/Mass of $C_{16}H_{34}$ = $16 \times 44/226$
= 3.12 kg of CO_2/1 kg of $C_{16}H_{34}$

Specific carbon intensity = Mass of carbon in fuel/calorific value of fuel
= $(12/44) \times 3.12 \times 1000/43$ = 19.79 g of carbon/MJ of diesel

Specific CO_2 intensity = Mass of CO_2/calorific value of fuel
= $3.12 \times 1000/43$ = 72.56 g of CO_2/MJ of diesel

$$C_2H_6O + 3(O_2 + 3.76N_2) \rightarrow 2CO_2 + 3H_2O + 11.28N_2 \tag{9.18}$$

Specific CO_2 molar concentration = Number of CO_2 moles/Number of $C_{16}H_{34}$
= 2/1 = 2 moles of CO_2/1 mole of DME

Specific CO_2 mass = Mass of CO_2/mass of $C_{16}H_{34}$ = $2 \times 44/46$
= 1.91 kg of CO_2/1 kg of DME

Specific carbon intensity = Mass of carbon in fuel/calorific value of fuel
= $(12/44) \times 1.91 \times 1000/43$ = 12.11 g of carbon/MJ of diesel

$$\text{Specific } CO_2 \text{ intensity} = \text{Mass of } CO_2 / \text{calorific value of fuel}$$
$$= 1.91 \times 1000/28 = 68.21 \text{ g of } CO_2 / \text{MJ of diesel}$$

$$\text{Percent change in } CO_2 \text{ emission} = (CO_2 \text{ from diesel} - CO_2 \text{ from DME})$$
$$/CO_2 \text{ from diesel} \times 100$$
$$= (72.56 - 68.21)/72.56 \times 100 = 6\%$$

Inference

If the fuel changes from diesel to DME, an internal combustion engine theoretically emits CO_2 emission that is 6% less with DME fuel than with base diesel.

4. Landfill gas fuel generates 20,000 SCM per day. The gas contains 60% methane and 40% carbon dioxide by volume. The gas is utilized by a spark-ignition engine with 30% thermal efficiency for 10 hours operation per day. Calculate the following:

   a. Potential of electric power generation using the gas per day

   b. CO_2 equivalent emission if the gas is utilized

   c. CO_2 equivalent emission if the gas is released into the open atmosphere

Solution

Input Parameters:

Volume of raw gas = 20,000 m³

Thermal efficiency, η = 30%

Duration of operation = 10 hours

Assumptions:

Calorific value of methane = 48 MJ/kg

Calorific value of CO_2 = 0 MJ/kg

Density of methane = .65 kg/m³

Density of carbon dioxide = 1.98 kg/m³

Density of raw gas = 1.182 kg/m³

Calorific value of raw gas = 28.8 MJ/kg

1. Mass flow rate $m_f = 20,000 \times 1.182/10 \times 3600 = 0.656$ kg/s

$$\text{Power available} = \eta \times m_f \times CV = 0.3 \times .656 \times 28.8 = 5.66 \text{ MW}$$

2. If the gas is utilized then it will combust as follows:

$$CH_4 + 2O_2 = CO_2 + 2H_2O$$

Available methane in the raw gas (by volume) = 60% of 20,000 = 12,000 m³

Available methane in the raw gas (by mass) = volume of methane × density of methane = 12,000 × .65 = 7800 kg methane

16 kg methane produces 44 kg carbon dioxide

7800 kg will produce 44 × 7800/16 = 21,450 kg CO_2

3. If the gas is released into the atmosphere then total CO_2 emission will be: CO_2 by combustion with methane + CO_2 in the raw gas = total CO_2 emission

$$21,450 + (40\% \text{ of } 20,000 \times 1.98) = 37,290 \text{ kg } CO_2$$

5. A transportation fleet in a city consumes 800 tonnes of CNG per day. If the following proposed fuels are to be introduced into the transportation sectors in the city, calculate the percent of change in CO_2 emission (local and net). The density of CNG and hydrogen are 0.8 and 0.07 kg/m³, respectively. The calorific value of CNG and hydrogen are 48 and 120 MJ/kg, respectively.

a. Hydrogen-blended natural gas (18% hydrogen by volume)

b. Ten percent of the fleet is to be battery-operated vehicles (coal power plant with 36% thermal efficiency, 27% transmissions and distribution (T&D) loss and 80% battery vehicle efficiency)

Solution

Parameters:

CNG requirement per day = 800 tonnes

Density of CNG = 0.8 kg/m³

Density of H_2 = 0.07 kg/m³

Calorific value of CNG = 48 MJ/kg

Calorific value of hydrogen = 120 MJ/kg

Assumptions:

CNG is assumed to have 100% methane.

1. If the entire fleet is converted to HCNG:

1 m³ of HCNG contains 0.18 m³ of H_2 and 0.82 m³ of CNG (CH_4)

Energy share of hydrogen in the blend:

$$E_h = 0.18 \times 0.07 \times 120 = 1.512 \text{ MJ}$$

Energy share of CNG in the blend

$$E_c = 0.82 \times 0.8 \times 48 = 31.48 \text{ MJ}$$

% of energy share by hydrogen: 1.512/(1.512 + 31.48) × 100% = 4.5%

$$CH_4 + 2O_2 = CO_2 + 2H_2O$$

In terms of energy:

$$16 \times 48 \text{ MJ methane produces } 44 \text{ kg of } CO_2$$

$$800 \times 48 \text{ MJ methane will produce } (44 \times 800 \times 48)/(16 \times 48) \text{ tonnes of } CO_2 = 2200 \text{ tonnes}$$

By using a HCNG blend, 4.5% equivalent energy will not produce any CO_2

Therefore, CO_2 produced by 4.5% of 2200 (= 99 tonnes) will not be emitted

2. If 10% of the fleet is converted to battery-operated vehicles, then the other 90% will produce CO_2:

$$0.9 \times 2200 = 1980 \text{ tonnes } CO_2$$

10% CNG (= 80 tonnes) is replaced by electricity from coal based power plant

CO_2 emission in power plant will take place from 80 tonnes of coal

$$C + O_2 = CO_2$$

$$44 \times 80/12 = 293.33 \text{ tonnes}$$

Overall efficiency of power availability at the user's end = $0.36 \times (1 - 0.27) \times 0.8 \times 100 = 20\%$

Therefore, the increase in global pollution will be $293.33 \times 0.21 = 61.59$ tonnes

Total CO_2 produced in this case will be $1980 + (293.33 + 61.59) = 2334.92$ tonnes

---

Unsolved Numerical Problems

1. An entire transportation fleet in a city uses butane as fuel. The fleet consumes about 1000 tonnes of fuel per day. Calculate the mass of CO_2 emission from the fleet. Assume that carbon-based emissions from the fleet are negligible.

2. 10,000 SCM of biogas in an industrial company is released into the atmosphere. An industrial company is planning to use the gas in an internal combustion engine for electrical power generation and the power is to be used in the plant itself. The biogas consists of 60% CH_4 and 40% CO_2. The thermal efficiency of the engine is 30% and the calorific value of the biogas is 28 MJ/kg. Calculate the potential power generation and CO_2 emission from the engine.

3. Calculate the CO_2 equivalent GHG emissions from a compression-ignition engine for 20-, 50-, and 200-year horizons. The engine emits 8% CO_2, 60 ppm N_2O, and 1500 ppm CH_4. The exhaust gas flow rate of the engine is 20 kg/h.

4. A methane-fueled internal combustion engine (rate power output 10 kW) emits 3 kg/hr. As brake thermal efficiency of the engine is now improved and the engine maintains the same power output of 10 kW, the engine now emits 2.75 kg/hr.

Calculate the change in thermal efficiency of the engine. The calorific value of methane is 48 MJ/kg and the combustion efficiency of the engine is 100%.

5. A 100-MW natural gas fueled gas turbine power plant has a thermal efficiency of 39%. The transmission and distribution loss is 30% If the same power is produced using a natural gas fueled internal combustion engine with a thermal efficiency of 35% and losses are nil, calculate the change in the specific fuel consumption and specific CO_2 emission for both power plants.

6. A biodiesel- ($C_{18}H_{22}O_3$) fueled diesel engine emits 700 ppm of CO, 50 ppm of HC, 8% of CO_2, and 0.01% of soot. Calculate the specific CO_2 emission (mole of CO_2/ mole of biodiesel).

References

Aslam, M. U., Masjuki, H. H., Kalam, M. A., Abdesselam, H. et al. (2006), An experimental investigation of CNG as an alternative fuel for a retrofitted gasoline vehicle, *Fuel*, 85(5–6), 717–724.

Behçet, R. (2011), Performance and emission study of waste anchovy fish biodiesel in a diesel engine, *Fuel Processing Technology*, 92(6), 1187–1194.

Bose, P. K., Deb, M., Banerjee, R. and Majumder, A. (2013), Multi objective optimization of performance parameters of a single cylinder diesel engine running with hydrogen using a Taguchi-fuzzy based approach, *Energy*, 63, 375–386.

Can, Ö. (2014), Combustion characteristics, performance and exhaust emissions of a diesel engine fueled with a waste cooking oil biodiesel mixture, *Energy Conversion and Management*, 87, 676–686.

Caton, J. A., Narney, J. K., Cariappa, H. C. and Laster, W. R. (1995), The selective non-catalytic reduction of nitric oxide using ammonia at up to 15% oxygen, *The Canadian Journal of Chemical Engineering*, 73(3), 345–350.

Çelik, M. B., Bülent Ö., Alkan, F. (2011), The use of pure methanol as fuel at high compression ratio in a single cylinder gasoline engine, *Fuel*, 90(4), 1591–1598.

Chintala, V. (2016), Experimental investigation on utilization of hydrogen in a compression ignition engine under dual fuel mode, PhD thesis, IIT Delhi.

Çinar, C., Şahin, F., Can, Ö. and Uyumaz, A. (2016), A comparison of performance and exhaust emissions with different valve lift profiles between gasoline and LPG fuels in a SI engine, *Applied Thermal Engineering*, 107, 1261–1268.

Department of Energy & Climate Change. www.gov.uk/government/uploads/system/uploads /attachment_data/file/295968/20140327_2013_UK_Greenhouse_Gas_Emissions_Provisional _Figures.pdf, accessed April 8, 2014.

EPA, United States Environmental Protection Agency, Global greenhouse gas emissions data, www .epa.gov/climatechange/ghgemissions/global.html, accessed April 6, 2014.

Federal Register, Vol. 76, No.179, September 15, 2011, http://www.gpo.gov/fdsys/pkg/FR-2011-09 -15/pdf/-20740.pdf, accessed April 8, 2014.

Feng, B., Liu, H., Yuan, J.-W., Lin, Z.-J. et al. (1996), Mechanisms of N_2O formation from char combustion, *Energy & Fuels*, 10(1), 203–208.

Gatts, T., Liu, S., Liew, C., Ralston, B. et al. (2012), An experimental investigation of incomplete combustion of gaseous fuels of a heavy-duty diesel engine supplemented with hydrogen and natural gas, *International Journal of Hydrogen Energy*, 37(9), 7848–7859.

Graham, L. A., Rideout, G., Rosenblatt, D. and Hendren, J. (2008), Greenhouse gas emissions from heavy-duty vehicles, *Atmospheric Environment*, 42(19), 4665–4681.

Gumus M. and Kasifoglu, S. (2010), Performance and emission evaluation of a compression ignition engine using a biodiesel (apricot seed kernel oil methyl ester) and its blends with diesel fuel, *Biomass and Bioenergy*, 34(1), 134–139.

Hassaneen, A., Munack, A., Ruschel, Y., Schroeder, O. et al. (2012), Fuel economy and emission characteristics of gas-to-liquid (GTL) and rapeseed methyl ester (RME) as alternative fuels for diesel engines, *Fuel*, 97, 125–130.

Hayhurst, A. N. and Lawrence, A. D. (1992), Emissions of nitrous oxide from combustion sources, *Progress in Energy Combustion and Science*, 18, 529–552.

He, B.-Q., Shuai, S.-J., Wang, J.-X. and He, H. (2003), The effect of ethanol blended diesel fuels on emissions from a diesel engine, *Atmospheric Environment*, 37(35), 4965–4971.

Heywood, J. B. (1988), *Internal Combustion Engines Fundamentals*, New York: McGraw-Hill, Inc.

IEA, World Energy Outlook 2008.

IPCC, Fourth Assessment Report: Climate Change 2007, Direct Global Warming Potentials.

Kahraman, E., Ozcanlı, S. C. and Ozerdem, B. (2007), An experimental study on performance and emission characteristics of a hydrogen fuelled spark ignition engine, *International Journal of Hydrogen Energy*, 32(12), 2066–2072.

Kalargaris, I., Tian, G. and Gu, S. (2017), Combustion, performance and emission analysis of a DI diesel engine using plastic pyrolysis oil, *Fuel Processing Technology*, 157, 108–115.

Korakianitis, T., Namasivayam, A. M. and Crookes, R. J. (2010), Hydrogen dual-fuelling of compression ignition engines with emulsified biodiesel as pilot fuel, *International Journal of Hydrogen Energy*, 35(24), 13329–13344.

Liu, G., Song, X., Yu, J. and Qian, X. (2002), AM1 study of the reaction between NCO and NO2 yielding N_2O and CO_2, *Journal of Molecular Structure: THEOCHEM*, 578(1–3), 93–98.

Lounici, M. S., Loubar, K., Tarabet, L., Balistrou, M. et al. (2014), Towards improvement of natural gas-diesel dual fuel mode: An experimental investigation on performance and exhaust emissions, *Energy*, 64(0), 200–211.

Lu, Z.-M. and Lu, J.-D. (2009), Influences of O_2 concentration on NO reduction and N_2O formation in thermal deNOx process, *Combustion and Flame*, 156(6), 1303–1315.

Maghbouli, A., Yang, W., An, H., Shafee, S. et al. (2014), Modeling knocking combustion in hydrogen assisted compression ignition diesel engines, *Energy*, 76, 768–779.

Miyamoto, T., Hasegawa, H., Mikami, M. and Kojima, N. et al. (2011), Effect of hydrogen addition to intake gas on combustion and exhaust emission characteristics of a diesel engine, *International Journal of Hydrogen Energy*, 36(20), 13138–13149.

Na, K., Biswas, S., Robertson, W., Sahay, K. et al. (2015), Impact of biodiesel and renewable diesel on emissions of regulated pollutants and greenhouse gases on a 2000 heavy duty diesel truck, *Atmospheric Environment*, 107, 307–314.

National Aeronautics and Space Administration (NASA), The current and future consequences of global change, Government of the United States Climate Change, vital signs of the planet: effects, http://climate.nasa.gov/effects/, accessed December 8, 2014.

Ozsezen, A. N. and Canakci, M. (2011), Determination of performance and combustion characteristics of a diesel engine fueled with canola and waste palm oil methyl esters, *Energy Conversion and Management*, 52(1), 108–116.

Ozsezen, A. N., Canakci, M., Turkcan, A. and Sayin, C. (2009), Performance and combustion characteristics of a DI diesel engine fueled with waste palm oil and canola oil methyl esters, *Fuel*, 88(4), 629–636.

Przybyla, G., Hadavi, S., Li, H. and Andrews, G. E. (2013), Real world diesel engine greenhouse gas emissions for diesel fuel and B100, SAE International Technical Paper 2013-01-1514.

Radojevic, M. (1998), Reduction of nitrogen oxides in flue gases, *Environmental Pollution*, 102(1, Supplement 1), 685–689.

Raheman, H. and Kumari, S. (2014), Combustion characteristics and emissions of a compression ignition engine using emulsified jatropha biodiesel blend, *Biosystems Engineering*, 123, 29–39.

Ramalingam, S., Rajendran, S. and Ganesan, P. (2016), Improving the performance is better and emission reductions from Annona biodiesel operated diesel engine using 1,4-dioxane fuel additive, *Fuel*, 185, 804–809.

Sahoo, P. K., Das, L. M., Babu, M. K. G. and Naik, S. N. (2007), Biodiesel development from high acid value polanga seed oil and performance evaluation in a CI engine, *Fuel*, 86(3), 448–454.

Sakthivel, G., Nagarajan, G., Ilangkumaran, M. and Gaikwad, A. B. (2014), Comparative analysis of performance, emission and combustion parameters of diesel engine fuelled with ethyl ester of fish oil and its diesel blends, *Fuel*, 132, 116–124.

Sheldon, A. E. and Williamson, S. (2005), Comparative assessment of hybrid electric and fuel cell vehicles based on comprehensive well-to-wheels efficiency analysis, *IEEE Transactions on Vehicular Technology*, 54(3), 856–862.

Specht, B. A. (1998), Synthesis of methanol from biomass/CO_2 resources, in 4th International Conference on Greenhouse Gas Control Technologies, Amsterdam.

Srinivasan, K. K., Mago, P. J. and Krishnan, S. R. (2010), Analysis of exhaust waste heat recovery from a dual fuel low temperature combustion engine using an organic rankine cycle, *Energy*, 35(6), 2387–2399.

Subramanian, K. A. and Chintala, V. (2013), Reduction of GHGs emissions in a biodiesel fueled diesel engine using hydrogen, ASME 2013 Internal Combustion Engine Fall Technical Conference, October 13–16, Dearborn, MI, Paper No. ICEF2013-19133, doi:10.1115/ICEF2013-19133; 2013.

Subramanian, K. A., Das, L. M. and Babu, M. K. G. (2008), Control of GHG emissions from transport vehicles: Issues & challenges, SAE International 2008, SAE Technical Paper No. 2008-28-0056.

Troy R. L., Semelsberger A. (2006), Dimethyl ether (DME) as an alternative fuel, *Journal of Power Sources*, 156(2), 497–511.

UK Greenhouse Gas Emissions, Provisional Figures and 2012 UK Greenhouse Gas Emissions, Final Figures by Fuel Type and End-User, 27 March 2014: Department of Energy & Climate Change 2013, www.gov.uk/government/uploads/system/uploads/attachment _data/file/295968/20140327_2013_UK_Greenhouse_Gas_Emissions_Provisional_Figures .pdf, accessed April 8, 2014.

Uma, R., Kandpal, T. C. and Kishore, V. V. N. (2004), Emission characteristics of an electricity generation system in diesel alone and dual fuel modes, *Biomass and Bioenergy*, 27(2), 195–203.

Yüksel, F. and Yüksel, B. (2004), The use of ethanol–gasoline blend as a fuel in an SI engine, *Renewable Energy*, 29(7), 1181–1191.

Yusaf, T. F., Buttsworth, D. R., Saleh, K. H. and Yousif, B. F. (2010), CNG-diesel engine performance and exhaust emission analysis with the aid of artificial neural network, *Applied Energy*, 87(5), 1661–1669.

Zheng, J. and Caton, J. A. (2012), Second law analysis of a low temperature combustion diesel engine: Effect of injection timing and exhaust gas recirculation, *Energy*, 38(1), 78–84.

10

Answers to Frequently Asked Questions

Answers to some commonly asked questions are compiled in this chapter.

1. What is the meaning of bio, biomass, and biofuel?

    The word bio is derived from bios, the Greek term indicating life. Biomass is an organic matter that has stored energy through the process of photosynthesis. Biofuel is a fuel derived from living matter (biomass). The sun's energy is converted into bioenergy (biomass) stored in plants (biomass) through the photosynthesis process by a biological reaction between water and carbon dioxide.

2. What are the similarities and differences between first-generation biofuels and second-generation biofuels?

    The main similarity between first- and second-generation biofuels is that both are produced from land in the lithosphere. The difference between them is that first-generation biofuels are produced from the primary crops, whereas second-generation biofuels are produced from biomass waste. For example, ethanol produced from sugar is a first-generation biofuel, whereas ethanol produced from waste molasses is a second-generation biofuel.

3. What are third-generation biofuels?

    Third-generation biofuels are produced from species that grow in the hydrosphere. For example, algae grows in an aquatic system.

4. Do fourth-generation biofuels exit?

    Generation indicates where the root of the biospecies comes from. First- and second-generation biofuels come from the lithosphere and third-generation biofuels come from the hydrosphere. Fourth-generation biofuels may come from the atmosphere. Womack et al. (2010) reported that the atmosphere is a habitat for microorganisms. Aerobiology deals with the biogeography of these species. The temperature of ambient air varies over Earth's surface until the end of the troposphere and stratosphere. Many microorganisms are capable of growth at temperatures near and below $0°C$ (Morita, 1975) and the organisms can be metabolically active at temperatures as low as $-18°C$ (Rothschild and Mancinelli, 2001). Biofuels may be produced from these species but this is currently only a hypothetical concept. Another scenario is that synthetic biology could deal with the direct conversion of solar energy to a fuel through engineered photosynthetic microorganisms using sunlight, water, and CO (Aro, 2015). Algae can grow in an artificial environment (artificial light, CO_2, H_2O, and nutrients) but as detailed above, this is already known as a third-generation biofuel. Therefore, fourth-generation biofuels are yet to be identified.

5. What is a carbon-neutral fuel?

    Carbon, which is fixed by a producer from the atmosphere and then released back into the atmosphere, is a carbon-neutral fuel. For example, a tree is grown by fixing carbon from carbon dioxide. If the tree gets fire due to natural and environmental causes such as lightening, fire due to rolling of stones etc., all the carbon stored in the tree is released back into the atmosphere. If the tree gets fire due to natural and environment causes such as lightening, fire due to rolling of stones etc., carbon-neutral may be defined as the difference between fixing and releasing carbon into the atmosphere is zero.

6. Are all fuels carbon-neutral?

   No. If fossil-based embodied energy is used for the processing of biomass (for upgradation of fuel quality), storage, transportation, dispensing, and so forth, the net carbon (carbon store in biomass ≠ carbon released to the atmosphere) will not be zero, and the degree of carbon neutral-fuel may not be zero.

7. What is the difference between renewable fuel and biofuel?

   The difference is the source of energy for its production. For example, if hydrogen is produced through an electrolyzer that receives electrical energy from a renewable source (solar or wind) system, it is called a renewable hydrogen fuel. If the hydrogen is produced from biomass feedstock, it is called a biohydrogen. Renewable fuel is not a biofuel because solar energy and wind are abiotic substances. However, biofuels can be considered renewable fuels.

8. What is the present level of CO_2 concentration in the atmosphere? Is it higher or lower when compared to earlier centuries?

   CO_2 concentration in the atmosphere as of January 2017 was 405.92 ppm (NASA). The concentration was lower in earlier centuries and was below 300 ppm until 1950. CO_2 concentration has been fluctuating for 400,000 years in the range of 150 to 300 ppm.

9. Why does CO_2 concentration increase?

   CO_2 emission in the atmosphere increased steeply due to increasing energy usage in industries. The GDP from industries is strongly related to energy consumption. If the GDP of a country increases, energy consumption will relatively increase, and the level of emission concentrations, including CO_2, will increase in the atmosphere. The IPCC confirms that anthropogenic GHGs, which have increased since the preindustrial era driven largely by economic and population growth, are now higher than ever. Therefore, the use of renewable energy systems is an option that would mitigate some of these problems.

10. Why was the CO_2 concentration above 150 ppm even before the industrial era (before 1950)?

    Producers convert CO_2 and H_2O into cellular biomass through the photosynthesis process under sunlight and CO_2 concentration decreases in the atmosphere. However, CO_2 concentration in the atmosphere increases due to forest fires, volcano eruption, its release from oceans, and carbon stored in rocks/earth crust. This is how nature adjusts the carbon balance in the environment, and interestingly, the concentration fluctuated between 150 ppm to 300 ppm (approximately).

11. Why does Earth warm? What are the serious consequences with a warming Earth?

    Greenhouse gases, such as carbon dioxide (CO_2), methane (CH_4), nitrous oxide (N_2O), perfluorocarbons (PFCs), hydrofluorocarbons (HFCs), and sulfur hexafluoride (SF_6), trap and retain the sun's energy while the sun's radiation reaches the atmosphere. Some of the sun's radiation reflected by Earth's surface are also trapped and retained by the greenhouse gases. Therefore, the average temperature of Earth increases. Global warming leads to climate change, resulting in glacier melting, drought, sea level rise, extinction of some species due to temperature rise, habitat loss, flood, and so forth.

12. Who should we blame for these consequences?

No one should be blamed! Humans need homes, but today we live in mostly artificial habitat systems because of deficiencies in the human adaptation system (the absence of special features). For example, people who live in hot places need artificial cooling of their habitats (homes) from air-conditioning systems, but camels in a hot desert environment do not need artificial cooling because they have adapted to withstanding high temperatures. Similarly, people living in colder regions need artificial heat but polar bears in arctic regions do not due to their adaptation. As well, food production and other essential commodities have to be increased with an increasing world population. The modern civilized world needs huge amount of energy for realizing these needs; however, energy systems emit harmful emissions into the atmosphere. It is understood that the quality of human life in terms of material comforts has improved, but we face serious consequences as emissions from these energy systems negatively affect our ecosystem.

13. How can CO_2 emission from energy systems be controlled?

There are several ways to reduce CO_2 emission from energy systems, such as improvement of energy efficiency, use of less or no carbon-content fuel, employing carbon capture and storage (CCS) systems and use of renewable energy systems.

14. What is the projected CO_2 concentration at the end of this century?

If present energy systems continue as they are, the CO_2 concentration in the atmosphere at the end of this century may reach 1000 ppm (The Royal Society, 2009). The pH level of ocean water may be less than 7.8 due to the higher concentration of CO_2 in the atmosphere. If coral reefs are lost due to low pH, a vital habitat will be lost, and the marine ecosystem will be severely affected.

15. If it is assumed that all energy systems are supposed to be converted completely to renewable energy systems and these systems emit zero CO_2 emission, would the CO_2 emission quickly return to the desired level of less than 300 ppm (reference year 1950)?

No, the CO_2 concentration in the atmosphere may not decrease as the ocean, which is another sink for CO_2, will release back its stored CO_2 to the atmosphere. As a hypothetical case, it becomes obvious that these energy systems do not emit any emissions into the atmosphere. Nature may take several centuries to stabilize the CO_2 by itself.

16. Does bioenergy lead to a conflict of food versus fuel?

Yes, that may occur. For example, some countries use sunflower oil as a feedstock for biodiesel production but some countries use it for edible purposes. If a food crop is converted to an energy crop, food production may be reduced. But today these factors are now taken into account for avoiding this type of conflict.

17. How is the food versus fuel conflict avoided?

Second- and third-generation biofuels would not affect food sources. Land that is declared by governments as not suitable for agriculture, which is called degraded land, can be used for energy crops. For example, *Jatropha curcas, Pongamia pinnata*, and other such crops, can grow in degraded land. If degraded land is not used for any purpose, it may become barren land. If the land is used for bioenergy crops, it preserves the ecosystem by generating oxygen, reducing soil erosion, creating habitats (at least for scavengers), and so forth. The food matrix may increase or decrease depending on the types of bioplants.

18. What are the criteria for selecting a bioenergy system over a solar energy system?

If land is declared as not suitable for agriculture crops and energy crops, the land may be considered for installation of a solar photovoltaic (P-V) system, or a solar thermal or wind energy system. If wind velocity is generally more than 4 m/s, the land is suitable for a wind energy system. Depending on the solar energy potential (solar irradiation) and atmospheric temperature, a solar P-V or solar thermal system may be selected. Certain land is banned for the use of renewable energy system, such as endangered species habitats, seismic regions, and tribal villages. Bioenergy may be allowed as an exception because it may provide positive effects, such as providing a habitat for endangered species or economic security to a tribal village.

19. Does *Jatropha curcas* affect biodiversity of an ecosystem? What are the effects of a *Jatropha curcas* plant on the environment and ecosystems?

The answer is uncertain as the benefit is based on what type of land is selected for cultivating the plant. If the *Jatropha curcas* plant is not carefully planned for or does not go through an environment audit, that may partly affect the biodiversity of the ecosystem. Herbivores/scavengers cannot consume the fruit and leaves of *Jatropha curcas* due to their poisonous nature. Spices such as *Pteridium aquilinum* and *Prosopis juliflora* may spread at a much faster rate and may not allow other spices to grow alongside them, which may negatively affect biodiversity and result in a decreasing food matrix in a particular ecosystem (e.g., forest ecosystem). If there is less biodiversity, there will be less or no diversity of species due to the absence of a food matrix. On the positive side, *Jatropha curcas* would at least enhance a habitat system that would otherwise become barren land. Thus, an ecosystem will not be affected if *Jatropha curcas* is planted in a degraded land.

20. If CO_2 concentration is zero, will an ecosystem improve?

No, CO_2 concentration cannot/should not be zero because producers need CO_2 for their photosynthesis processes. If producers do not survive, herbivores, carnivores, and scavengers cannot live. Therefore, achieving zero CO_2 concentration cannot not be a desirable goal.

21. Does a hydrogen energy system achieve zero CO_2 emission?

If hydrogen fuel is produced from a renewable energy system (solar, wind, etc.), a heat engine will not emit CO_2 emission. However, if hydrogen is produced from a fossil fuel (natural gas, oil, etc.), localized emission from the engine will be zero but CO_2 from the fuel refineries would emit CO_2 emission.

22. Which system would give beneficial results: a natural gas fueled transportation fleet or a hydrogen- (produced from natural gas) fueled fleet?

CO_2 emission from a natural gas fueled fleet is less than hydrogen produced from natural gas refineries because the conversion efficiency of hydrogen is less than 100%. In addition, more energy input is required for compressing, storing, and transporting hydrogen. Therefore, CO_2 emission is high. The advantage with hydrogen is that local urban pollution would be less; however, global pollution would increase. If natural gas is converted to hydrogen with a CCS system, a hydrogen-fueled fleet is more beneficial in terms of zero carbon-based emissions (CO, CO_2, HC, smoke/PM, soot, etc.).

23. Why does CO_2 decrease with an increasing energy efficiency?

If the energy efficiency of a system increases, its energy consumption by that system will decrease, and correspondingly, emission will also decrease.

24. Why is it not feasible to adopt a CCS system in automotive vehicles?

A CCS system in the mobility system increases payload of a vehicle due to its additional weight and volume, resulting in an energy penalty. The captured CO_2 gas needs extremely large tanks for its storage, and therefore, from a practical standpoint, a CCS system is not feasible for automobile vehicles.

25. Do biofueled vehicles emit CO_2 emission?

Yes, biofueled vehicles emit CO_2 emission. Biomass is composed of carbon, hydrogen, oxygen, and so forth, and therefore, a vehicle fueled with biofuels (except biohydrogen) would emit CO_2 emission. For hydrogen, CO_2 emission will come from the biorefineries, but CO_2 can be recycled by the plants.

26. Why is bioenergy preferable to renewable energy?

A producer gives us biomass with tangible benefits such as oxygen release to the atmosphere, natural CO_2 sequestration, habitats for organisms, function of natural cycles, enhancement of soil quality by nitrate (fixing of nitrogen to nitrate by certain bacteria), and preventing soil erosion. These are just some of the benefits while implementing biofuels. Even though solar or wind energy systems do not emit the emissions, it may affect an ecosystem if it is installed in an undesirable place. For example, if too many kilometers are occupied by a solar P-V system, this may affect the habitat of organisms, including scavengers, resulting in severe damage to the ecosystem. High noise generated by a wind system may disturb certain organisms who may be forced to leave their habitat. But if these systems put in desert land or degraded land where ecosystems do not exit, these renewable energy systems will yield desirable results. Therefore, careful planning of renewable systems is needed; otherwise, they may negatively affect the ecosystems.

27. Is sulfur-based emission lower with biofueled internal combustion engines?

Yes. The sulfur content in biofuels is lower compared to fossil fuels (coal, oil, and natural gas). Sulfur is embedded with hydrocarbon fuel under the earth, whereas biomass grows over the earth with little or no sulfur.

28. Why does CO emission form in internal combustion engines?

Carbon monoxide (CO) forms due to a rich air-fuel mixture, poor air-fuel mixing (global mixture is less but local mixture is rich), and lower in-cylinder temperature. In addition, CO also forms at high temperature due to dissociation of CO_2 emission.

29. Why does NOx form in internal combustion engines?

Oxides of nitrogen (NOx) form when nitrogen reacts with oxygen at high temperatures (called thermal NOx). It is mainly a function of temperature, oxygen concentration, and reaction time. NOx can also form with flame by HCN radical (prompt NOx) and fuel-bound nitrogen (fuel bound NOx).

30. Could an IC engine emit zero NOx emission?

    Yes, it is possible. When an IC engine is operated with 100% oxygen instead of air, the engine can emit zero NOx emission. However, if the fuel contains nitrogen, emission with an oxy-combustion will form. Note that the fuel for an IC engine contains almost none or negligible nitrogen embedded with hydrocarbon.

31. Could an alcohol- (methanol, ethanol, and butanol) fueled engine provide higher thermal efficiency for a spark-ignition engine compared to base gasoline?

    If the higher-octane-number alcohol fuel is used in a conventional (unmodified version) SI engine, the efficiency will not be higher. But if the compression ratio of the engine is increased and optimized, an engine fueled with alcohol fuel could give provide thermal efficiency.

32. Could raw biogas fuel give better results in terms of an engine's performance and emission reduction compared to enriched biogas?

    No. A raw biogas fueled engine cannot give the same or better performance and emission reduction as one that uses enriched biogas. This is mainly due to less flame velocity of raw biogas because of the CO_2 content in the fuel, resulting in poor combustion or delayed heat release. On the other hand, an advantage is that NOx emission with raw biogas is lower than that of enriched biogas due to dilution and a thermal effect from the CO_2 content in fuel. Performance can be improved by increasing the compression ratio but power drop will be a problem.

33. Are enriched biogas and fossil methane the same?

    Not exactly. The similarity is that both fuels have methane. However, other gas compositions are different. Enriched biogas primarily consists of methane (pp to 95%), and the remaining gases are nitrogen, CO_2, hydrogen, and H_2S. Enriched biogas contains primarily methane, and the remaining gases are ethane, propane, propane, butane, CO_2, and so forth. The specific gravity and calorific value of these two fuels are different and an engine needs fuel calibration for both fuels.

34. Is enriched biogas economical? Why cannot not raw biogas able to inject into natural gas pipeline?

    It may be economical if the enrichment plant operates at a large scale. As raw biogas contains about 40%–50% carbon dioxide, it is not generally preferable to store the gas in cylinders or to transport it through a pipeline due to increased failure rate of the cylinders caused by brittleness as well as the different composition of methane gas. The gas contains a lower calorific value than enriched biogas resulting in power drop in automotive vehicles. Water scrub and the membrane method are widely used for CO_2 removal in raw biogas. If the biogas needs to be injected into a natural gas grid (pipeline) or stored in cylinders for its utilization in automotive vehicles, enriched biogas is mandatory.

35. Is hydrogen viable to implement in internal combustion engines?

    The researcher reported that hydrogen use in internal combustion engines could give better performance and carbon-based emissions reduction due to its desirable fuel quality, including higher flame velocity and octane number. Its drawbacks are primarily storage and safety.

36. If the prices of 1 kg of gasoline and ethanol are 60 rupees and 45 rupees, respectively, is ethanol cheaper than gasoline?

No. the price of gasoline on an energy basis is lower (1.36 Rs/MJ (calorific value: 44 MJ/kg)) than ethanol (1.6 Rs/MJ (calorific value: 28 MJ/kg)). In this case, gasoline is cheaper than ethanol.

37. It is reported in most studies that the price of biofuel is generally higher than that of petroleum fuel. Is it economically feasible?

Biofuels may be economically feasible if they are produced on a large scale and have government support from subsidies, incentives, carbon credit, and so forth. Funds generated through a carbon tax on petroleum fuel can also boost a biofuels program. However, even if it may not be economical, it should not matter since biofuels are important in terms of maintaining a green and sustainable environment.

38. What is the difference between clean fuel and green fuel?

The difference between clean fuel and green fuel is based on what resource used for the fuel production. For example, if hydrogen is derived from fossil natural gas, it is called a clean fuel, whereas green fuel is when hydrogen is produced from renewable energy sources (solar or wind). A fuel that does not lead to deterioration of the environment at any point in time is a green fuel, whereas fuel that would lead to maintaining a relatively better or cleaner environment with reference to a base condition is a clean fuel. Another example would be that DME fuel derived from coal is a clean fuel (compared to coal), whereas DME derived from biomass is a green fuel.

39. What is the difference between soot and particulate matter?

Particulate matter is comprised of organic (carbon, hydrocarbon) and inorganic substances (sulfur, metals from lubricating oil, ketone, etc.). Soot is made of carbonaceous substance. Soot is in the category of particulate matter.

40. How does PM form in internal combustion engines?

PM forms in internal combustion engines due to rich mixture, poor mixing, lower or insufficient in-cylinder temperature, and inorganic substance.

41. Why does a gasoline-fueled spark-ignition engine emit less PM emission than a diesel-fueled CI engine?

Gasoline fuel will mix better due to high vaporization tendency and relatively more time availability (nearly half of the cycle time at the same engine speed) for mixture preparation process. Combustion with a premixed gasoline charge leads to less PM emission. In the case of a CI engine, the globalized air-fuel mixture is rich but the localized mixture is rich. Diesel in a liquid state has less diffusivity than gasoline fuel, resulting in poor mixing of diesel with air. The mixture preparation time in a CI engine is relatively much less than in an SI engine. These are some of the reasons why a diesel-fueled engine emits a high level of PM emission.

42. Which is better, an SI engine or a CI engine?

Under the same conditions (e.g., compression ratio, speed, swept volume), an SI engine delivers a higher specific power output than a CI engine. If a CI engine can operate at a higher compression ratio, the thermal efficiency of a CI engine is higher than that of a SI engine.

43. Is torque a function of speed? Why does torque occur in the middle of engine speed, whereas maximum power occurs at high engine speed?

    Torque is mainly a function of mean effective pressure and swept volume and not a function of speed, as torque does not keep increasing with increasing speed. Maximum torque occurs relatively at a lower speed than the speed at which the engine delivers the rated power output. The reason is that the intake valve opening period decreases with increasing speed and hence at higher engine speed, lower volumetric efficiency leads to lower torque. Even though the torque is lower at high engine speed, the speed is dominant, resulting in higher power output.

44. Could the cetane number of fuel affect the life of compression-ignition engines?

    Yes. It can be related to the life of the engines. The cetane number of fuel indicates its ignition quality. If the cetane number is high, the startabilty of the engine is high, and thus will eliminate the chance of knock. Knock is defined as undesirable combustion phenomena that will tend to raise the rate of pressure, which may damage the engine components. Therefore, a high-cetane-number fuel can eliminate knock and damage to engine components, which may result in increasing engine life. A high cetane number fueled engine improves transient engine operation by reducing white smoke and other transient emissions.

45. Can a fuel have a high octane number as well as a high cetane number?

    No. A fuel has generally either a high octane number and low cetane number and vice versa.

46. Do biodiesel, DME, and F-T diesel fueled diesel engines have cold-start ability problems compared to those using base diesel?

    No. The cold-start ability problem will be less with these biofuels due to a higher cetane number than that of base diesel. The cetane number indicates ignition quality of fuel and ability to burn a fuel at relatively low temperatures.

47. Does a high cetane number mean high thermal efficiency?

    There are no consistent results showing that the cetane number of fuel increases with an increasing cetane number. As the cetane number increases, premixed combustion decreases due to less ignition delay, leading to a lower degree of constant volume combustion that may result in lower thermal efficiency. On the other hand, combustion duration would decrease with an increasing cetane number due to its effective burning, even relatively at low temperature, resulting in fast burning (due to less activation energy requirement) and higher thermal efficiency. The thermal efficiency of the engine is not only dependent on fuel quality but also on the type of engine technology.

48. Could the compression ratio of a compression-ignition engine increase with an increasing cetane number?

    A compression-ignition engine does not thermodynamically have limitations on increasing compression ratio. However, thermal stress on the engine components (cylinder liner, valves, and cylinder head) increases with an increasing compression ratio, resulting in affecting engine life. Therefore, large-sized engines (in MW range) have a relatively lower compression ratio compared to small-sized engines primarily due to enhancing the life of the engine. The surface-to-volume ratio increases with an increasing compression ratio and it may lead to high hydrocarbon emissions. The lubricating oil performance and wall heat transfer may be affected with increasing compression ratio. OEMs will decide the suitable compression ratio based on these factors. In general, it is well established that the thermal efficiency of a compression-ignition engine increases with an increasing compression ratio. In contrast, an engine with a lower compression ration can operate with a higher-cetane-number fuel without a knocking problem.

49. Does higher-octane-number fuel lead to an increase in the thermal efficiency of a spark-ignition engine?

If the octane number fuel is increased further but is used in a conventional SI engine (unmodified version), the thermal efficiency of an engine will not increase. However, if the compression ratio of an engine is increased and optimized and operated with a higher-octane-number fuel, the engine's thermal efficiency will be higher with a higher-octane-number fuel than with a base-octane-number fuel.

50. Which fuel—biodiesel, F-T, or DME—is a better choice for compression-ignition engines?

A DME-fueled engine could give nearly zero smoke and less PM emissions with almost negligible carbon deposits on pistons and the nozzle (coking), but it needs a customized fuel storage and handling system. In the case of F-T, better performance and emission reduction (due to negligible aromatic structure) can be achieved but PM count and number are a challenge compared to DME. In the case of biodiesel, cold flow, fuel storage stability, carbon deposit, lubricating oil dilution, and higher NOx emission are prime problems. DME could be a better choice from an emission point of view.

51. Which biofuel would be feasible to implement immediately in a current transportation fleet?

Diesel-blended F-T diesel/biodiesel can be immediately implemented in a fleet. These fuels do not need major modification of the fuel-handling infrastructure. DME needs a dedicated fuel-handling system, including storage, transportation, and dispensing.

52. If hydrogen is produced from (1) natural gas feedstock through the steam reforming method (SRM) or (2) water as feedstock through electrolyzer, is the efficiency of a hydrogen-fueled vehicle is same for both?

Yes, the engine efficiency of the vehicle, which is called the tank-to-wheel efficiency, is the same irrespective of the source and production route. But well-to-wheel (net life cycle efficiency) will not be the same. The well-to-tank efficiency of the vehicle fueled with hydrogen produced from the SRM method is higher than from the electrolyzer method. Therefore, the well-to-wheel efficiency of the first system is higher than that of the second system, and hence each systems' life cycle efficiency is different.

53. Do biofuels have superior fuel quality than that of fossil gasoline and diesel fuel?

In general, the answer is yes. The octane number of biofuels (alcohol, biomethane, and hydrogen) are higher than petrogasoline fuel, whereas the cetane number of biofuels (biodiesel, F-T diesel, DME) are generally higher than that of petroleum diesel. If the biofuels have any shortcomings in any of the other physicochemical properties, the engine parameters need to be optimized.

54. Could biofuel be utilized in unmodified engines? If not, how is there better performance and emission reduction of engines with biofuels?

No, it is not desirable to use biofuels (100%) in unmodified engines. An engine's design and operating parameters need to be optimized for biofuels in order to get better engine performance with less emissions. OEM-approved biofueled vehicles should always be used in terms of better human, vehicle, and environmental safety, as well as for getting the benefits, including high fuel economy and increased engine/vehicle life.

55. Biogas from a municipal waste or wastewater treatment plant is released naturally. The gas is trapped and utilized in an internal combustion engine, but the engine emits CO_2 emission (greenhouse gas). Which is the beneficial option: if biogas is liberated to the atmosphere as such, or CO_2 addition to the atmosphere after the biogas is utilized in an IC engine?

Biogas consists of approximately 60% methane and 40% CO_2. Methane gas has a higher global warming potential (25 in a 100-year horizon) than that of CO_2. Therefore, the methane in biogas is converted into power by an IC engine in which the reactant CH_4 is converted to product CO_2, and subsequently, it is added to the atmosphere. CO_2 can be synthesized by producers even it may take a few decades or centuries, but synthesizing of CH_4 into another product by natural conservation is a problem. Therefore, instead of releasing biogas, biogas utilization in internal combustion engines could provide positive results.

56. Could biofuels replace petroleum fuel?

As of now, biofuels are not able to replace petroleum fuel due to their limited availability because first- and second-generation biofuels are generated from the biomass grown in the lithosphere. Land availability in the lithosphere for energy crops is limited due to the land being occupied by forest, agriculture, natural water bodies, and mining, desert, industries, civil construction, and so forth. However, as two-thirds of Earth is covered by water (hydrosphere), third-generation biofuels (algae) would theoretically have a high potential and thus biofuel could meet the required fuel needs in the future.

57. What are the historical events regarding biofuel use in internal combustion engines?

Biofuel, such as alcohol-blended gasoline, was used in vehicles/military battle tanks during World War II (1939–1945) due to a serious shortage of conventional fuel (petroleum fuel). The great technocrat Nikolaus August Otto, who is the inventor of the spark-ignition engine, preferred the use of ethanol in an engine. Industrialist Henry Ford designed a vehicle (Model T) to run on ethanol (Henry Ford, www.thehenryford.org). Another great technocrat, Rudolf Diesel, the German inventor of the compression-ignition engine, designed his diesel engine to run with peanut vegetable oil and he demonstrated the engine at the World Exhibition in Paris in 1897 (Biofuel Facts, www.biofuel.org.uk).

58. How can we eliminate the food versus biofuel conflict?

There are many models available to eliminate the conflict. One model is that bioenergy/biofuel can be derived from waste biomass. For example, in a wheat paddy, wheat grain is a food for humans, wheat straw is for animals, and waste biomass (wheat husk) is used for biofuel production. Another model is that the electricity produced by a renewable energy system (wind/solar) is fed to a centralized electrical grid. If excess (surplus) electricity is available, it is used for an artificial bioreactor that synthesizes CO_2, water, and hydrogen into starch, biomass, biofuel, biochemical substances, and so forth under artificial lighting. Food is the main product and the waste biomass can be used for energy purposes. The biomass can grow at a faster rate (24/7) throughout the year through this artificial bioreactor. A future strategy has to focus on both food and fuel for sustainable development of a bioenergy system. In this way, the food versus fuel conflict can be eliminated. Biomass in open land is mandatory for increasing habitat area and the food web for herbivores, carnivores, and scavengers for maintaining a sustainable ecosystem. Biofuel/bioenergy could strengthen sustainable energy, the environment, and ecology systems.

References

Aro, E.-M. (2016), From first generation biofuels to advanced solar biofuels, *AMBIO*, 45(Suppl. 1), 24–31.

Biofuel facts, www.biofuel.org.uk, accessed February 26, 2017.

Climate change, 2014 synthesis report summary for policy maker, IPCC. www.ipcc.ch/pdf/assessment-report/ar5/syr/AR5_SYR_FINAL_SPM.pdf/, accessed February 26, 2017.

Morita, R. (1975), Psychrophilic bacteria, *Microbiology and Molecular Biology Reviews*, 39, 144–167.

NASA, Global climate change, http://climate.nasa.gov/vital-signs/carbon-dioxide/, accessed January 16, 2017.

Rothschild, L. J. and Mancinelli, R. L. (2001), Life in extreme environments, *Nature*, 409, 1092–1101.

The Henry Ford, The model T, https://www.thehenryford.org/collections-and-research/digital-collections/sets/737, accessed February 26, 2017.

The Royal Society (2009), *Geoengineering the Climate: Science, Governance and Uncertainty*, London: The Royal Society.

Womack, A. M., Bohannan, B. J. M. and Green, J. L. (2010), Biodiversity and biogeography of the atmosphere, *Philosophical Transactions of the Royal Society B*, 365, 3645–3653.

Index